This book is a comprehensive design text for permanent magnets and their application. Permanent magnets are very important industrially, and are widely used in a variety of applications, including industrial drives, consumer products, computers and automobiles.

In the early 1970s a new class of magnet – the rare earths – was discovered, the properties of which showed sustained improvement over the following two decades. New materials such as these have spawned many new markets for magnets, with significant performance gains in the devices for which they are used. By the early 1990s the new magnet technologies had matured. Until the advent of the present book, however, there has been no text that unified all the relevant information on the wide range of modern permanent magnet materials. This book, therefore, has been written as a comprehensive review of the technology, intended for scientists and engineers involved in all stages of the manufacture, design and use of magnets. A brief theory of magnetism explains the behavior of the different classes of permanent magnet, and the various production processes that lead to quite diverse material characteristics. The core of the book is a detailed treatment of the methods that are used to design permanent magnets, including assessments of the changes they experience under practical operating conditions. Modern analytical techniques are described, including the finite element method, with reference to the accurate simulation of permanent magnet materials. With the evolution of new materials, the markets for permanent magnets have changed. In this book, the author emphasizes the most important modern applications, and discusses the viability of the various magnet types that are now available.

This book, the first to cover comprehensively all aspects of modern permanent magnet materials, their design and application, will be of value to anyone involved in the design and use of magnets.

PERMANENT MAGNET MATERIALS AND THEIR APPLICATION

PERMANENT MAGNET MATERIALS AND THEIR APPLICATION

PETER CAMPBELL

President, Princeton Electro-Technology, Inc., Boca Raton, Florida

CAMBRIDGE UNIVERSITY PRESS
Cambridge, New York, Melbourne, Madrid, Cape Town, Singapore,
São Paulo, Delhi, Dubai, Tokyo, Mexico City

Cambridge University Press
The Edinburgh Building, Cambridge CB2 8RU, UK

Published in the United States of America by
Cambridge University Press, New York

www.cambridge.org
Information on this title: www.cambridge.org/9780521566889

First published 1994
First paperback edition 1996

A catalogue record for this publication is available from the British Library

Library of Congress Cataloguing in Publication Data

Campbell, Peter, 1949–
Permanent magnet materials and their application / Peter Campbell
p. cm.
Includes index.
ISBN 0 521 24996 1
1. Permanent magnets – Design and construction. 2. Magnetic
materials. I. Title.
QC757.9.C36 1994
621.34-dc20 93-43324 CIP

ISBN 978-0-521-24996-6 Hardback
ISBN 978-0-521-56688-9 Paperback

Contents

Preface

The original inspiration to write this book came when, after an electrical engineering training in the late 1960s, I embarked upon the design of a variety of permanent magnet electrical machines. I needed to know more about the behavior and performance of the different magnet materials than the electromechanical design texts provided, and significantly more applications data than the scientific books on magnetism contained. This shortcoming was exacerbated in the early 1970s when an entirely new class of magnet – *the rare earths* – was discovered, offering a vast array of new opportunities for permanent magnet devices, and new challenges to designers such as myself. As these new materials were developed, their properties exhibited dramatic improvements from year to year, reaching maturity in the early 1990s as a full range of samarium–cobalt and neodymium–iron–boron magnets. Until this had happened, I felt that any attempt to produce a comprehensive text including a description of these materials would have been premature. Now, with first-hand experience in most cases, I am able to describe their selection and design for a wide range of important applications.

The material for this book has evolved from courses given to students and practicing engineers while I was at the University of Cambridge and the University of Southern California, and from a variety of assignments to develop and design permanent magnet materials. Though this is intended mainly as a design text, I thought it important to open with an explanation of the theory of permanent magnetism, so the rationale for the various production processes can be understood, leading to the distinct properties of the different classes of magnet. Nevertheless, the core of this book is a detailed treatment of the methods that are used to design permanent magnets, but an important practical consideration is the changes that occur in material properties due to environmental conditions.

The difference between *design* and *analysis* is emphasized, and an appropriate simulation for a permanent magnet is described for each technique, including the popular finite element method.

With the addition of the rare earth class of magnet to the existing ferrite and alnico types, there are new devices, which represent significant new markets for magnets, and there are existing devices whose performance has been substantially improved by changing to a modern material. The selection and optimization of a permanent magnet for the most important applications is described in this book, but the list of products is noticably different from any that would have appeared prior to commercialization of the rare earths. My hope is that the information contained in this book will make manufacturers, designers and users aware of the very broad range of properties available from today's permanent magnets, and arm them with the skills necessary to develop more, successful, new products.

Los Angeles, California *Peter Campbell*

1
Fundamentals of magnetism

1.1 Introduction

Earlier texts on permanent magnets have opened with historical reviews of these materials (Hadfield, 1962; McCaig, 1977; Parker, 1990; Parker and Studders, 1962). In this book the design of modern permanent magnets is emphasized, and so initially the development of the relationships that are required to model today's materials for a variety of common applications is considered. To the extent that a historical review is provided in this chapter, it is of those fundamental equations of electromagnetism that are needed to understand the performance of magnets in circuits and devices.

There are many properties of a permanent magnet that are considered in its design for a magnetic device, but most often it is the *demagnetization curve* that initially determines its suitability for the task. Its shape contains information on how the magnet will behave under static and dynamic operating conditions, and in this sense the material characteristic will constrain what can be achieved in the device design.

The *B versus H* loop of any permanent magnet has some portions which are almost linear, and others that are highly non-linear. The shapes of these *B versus H* loops, or at least the *demagnetization* portions of them, tell the designer a lot about the suitability of the material for a given application. A brief derivation of the *B versus H* loop is presented, to illustrate the microscopic mechanisms that determine the macroscopic performance of a magnet.

The fundamental theory describing permanent magnets is similar to that for soft magnetic materials. A magnet is permanent (sometimes called *hard*) if it will alone support a useful flux in the air gap of a device, whereas it is *soft* if it can only do so with the aid of an external electrical or magnetic input. The most basic parameter of either type of material is the *magnetic dipole moment*.

1.2 Magnetic dipole moment

The magnetic dipole moment may be modeled in a similar manner to a loop of wire carrying a current i. Consider that this loop lies in the x–y plane shown in Figure 1.1, and that the loop is divided into strips such as *abcd*, which will each carry i (this is legitimate since the currents in sides *ab* and *cd* will cancel for neighboring strips). There is a magnetic field $\boldsymbol{B}$ (*flux density*) in the magnet, which we shall take to lie in an x–z plane. $\boldsymbol{B}$ may then be resolved into components $B \cos \theta$ in the z direction and $B \sin \theta$ in the x direction, perpendicular and parallel to the plane of the loop respectively. The force due to a magnetic field on a conductor carrying electrical current is known to be proportional to the current (i), flux density (B) and conductor length (l), and the constant of proportionality is the sine of the angle between $\boldsymbol{i}$ and $\boldsymbol{B}$. Thanks to our choice of orientation for the loop and the field, the forces in all four sides *ab*, *bc*, *cd* and *da* due to $B \cos \theta$ are in equilibrium, and the forces in the long sides *ab* and *cd* due to $B \sin \theta$ are zero. However, those in sides *bc* and *da* due to $B \sin \theta$ form a couple (or torque) of magnitude

$$\delta \boldsymbol{T} = abi\, \delta y\, B \sin \theta \tag{1.1}$$

Because the area of the strip is $\delta A = ab\, \delta y$, this becomes

$$\delta \boldsymbol{T} = i\, \delta A\, B \sin \theta \tag{1.2}$$

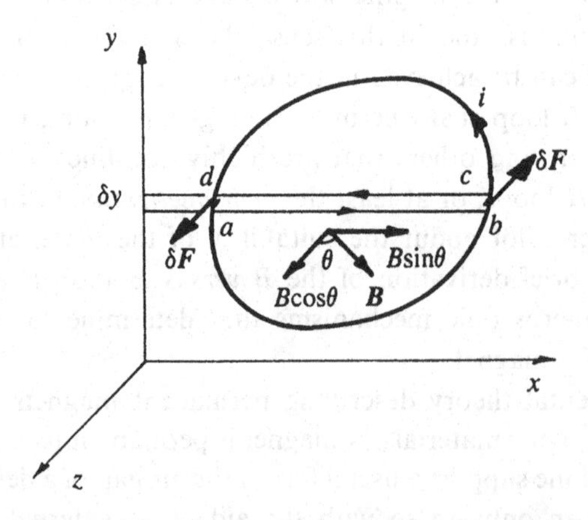

Figure 1.1. A loop carrying current i in field $\boldsymbol{B}$.

Even when modeled as a current loop, the magnetic dipole moment is still a microscopic property of the material. All the strips that comprise the loop in Figure 1.1 carry the same current i, and the entire loop will experience a unique field $\boldsymbol{B}$. The torque on the loop is therefore a summation of Equation (1.2) for all the $\delta \boldsymbol{A}$ areas that constitute the complete loop $\boldsymbol{A}$, so

$$\boldsymbol{T} = i\boldsymbol{A}B \sin\theta \tag{1.3}$$

The boundary of area $\boldsymbol{A}$ is coincident with the path of i, so it is natural to define their product as a unique parameter, which is the *magnetic dipole moment*:

$$\boldsymbol{\mu}_{\mathrm{m}} = i\boldsymbol{A} \tag{1.4}$$

The torque on the current loop now becomes

$$\boldsymbol{T} = \boldsymbol{\mu}_{\mathrm{m}} B \sin\theta \tag{1.5}$$

Torque $\boldsymbol{T}$, moment $\boldsymbol{\mu}_{\mathrm{m}}$ and the field $\boldsymbol{B}$ (as $B \sin\theta$) are all vector quantities with specific directions as well as magnitudes. No particular shape was defined for the loop in Figure 1.1, so the area $\boldsymbol{A}$ may describe any shape of loop provided that it is planar. The vector $\boldsymbol{A}$ is normal to that plane, so too will be $\boldsymbol{\mu}_{\mathrm{m}}$.

The rectangular current loop of Figure 1.2 has $\boldsymbol{\mu}_{\mathrm{m}}$ normal to its plane, and in the presence of a magnetic field $\boldsymbol{B}$ there is a torque $\boldsymbol{T}$ about its axis. If the loop can rotate about this axis, then the torque tries to align $\boldsymbol{\mu}_{\mathrm{m}}$ with the direction of $\boldsymbol{B}$; if this is achieved, then θ and $\sin\theta$ are zero, as will be $\boldsymbol{T}$.

An electron in a microscopic orbit has a magnetic dipole moment, which we have modeled as a current circulating in a loop. A simple model of a bulk magnetic material is a large array of electrons, and this material will appear to be unmagnetized if its magnetic dipole moments are randomly

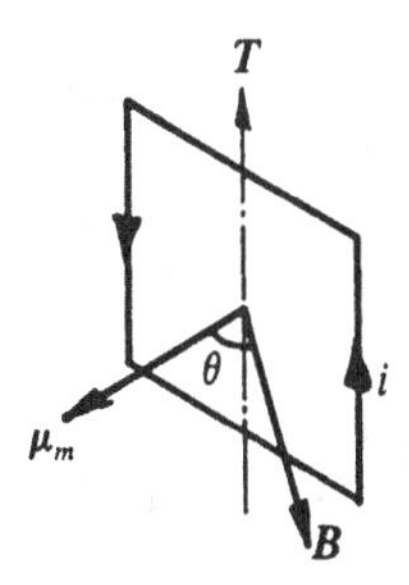

Figure 1.2. Torque applied to a magnetic dipole current loop.

oriented. When a field $\boldsymbol{B}$ is applied to the material, each moment experiences a torque that tends to rotate it towards the direction of $\boldsymbol{B}$. There are atomic forces on the electron orbits, which resist rotation by $\boldsymbol{T}$, but if the applied field is strong enough, then the torques will be great enough to align all the moments with $\boldsymbol{B}$. The material has then reached its *saturation* field. This mechanism is not sufficient, however, to create a permanent magnet material, which by our earlier definition must be able to sustain its own magnetic flux in the absence of any external sources of field. In other words, a permanent magnet must sustain flux by virtue of its own internal field, which will require spontaneous alignment of the magnetic dipole moments, or *spontaneous magnetization*.

The work done in rotating the moment in a field is found by integrating Equation (1.5) through the angle of rotation,

$$\begin{aligned} E &= \int \boldsymbol{T} \cdot \mathrm{d}\boldsymbol{\theta} \\ &= \mu_{\mathrm{m}} B \int \sin\theta \cdot \mathrm{d}\theta = -\mu_{\mathrm{m}} B \cos\theta \end{aligned} \tag{1.6}$$

The lowest value of energy E in the material occurs when $\boldsymbol{\mu}_{\mathrm{m}}$ and $\boldsymbol{B}$ are aligned, so there is indeed a natural tendency for this internal alignment to happen and hence minimize the energy. This process of aligning the axes of magnetic dipoles due to their own internal field is called *exchange interaction*. In an elemental volume of the bulk material, the adjacent values (and directions) of $\boldsymbol{\mu}_{\mathrm{m}}$ will be identical, so a summation of $\boldsymbol{\mu}_{\mathrm{m}}$ may be performed over this volume ΔV, which yields a new property of the material called its *magnetization* $\boldsymbol{M}$:

$$\boldsymbol{M} = \lim_{\Delta V \to 0} \frac{\sum \boldsymbol{\mu}_{\mathrm{m}}}{\Delta V} \tag{1.7}$$

This strict definition of $\boldsymbol{M}$ shows that it is actually the magnetic dipole moment per unit volume, but unlike $\boldsymbol{\mu}_{\mathrm{m}}$, it is a macroscopic property of the material, like $\boldsymbol{B}$. In an elemental volume alone, although the dipoles are aligned by the internal field, that field $\boldsymbol{B}$ will *itself* be generated by $\boldsymbol{M}$. The existence of an internal field even when no external field is applied is the phenomenon of *spontaneous magnetization*. This direct relationship is expressed as $\boldsymbol{B} = \mu_0 \boldsymbol{M}$ (where μ_0 is a constant of proportionality), which allows Equation (1.6) to be rewritten as

$$E = -\mu_0 \mu_{\mathrm{m}} M \cos\theta \tag{1.8}$$

The foregoing is clearly a simplified theory of magnetization, which neglects many of the practical conditions that occur. For example, as we shall see later, thermal agitation will disrupt the alignment of the moments and reduce their magnetization $\boldsymbol{M}$.

1.3 Magnetizing force

While spontaneous magnetization yields a specific value of $\boldsymbol{M}$ in an elemental volume of a magnet, it is not likely that an entire magnet will operate in the same condition with a unique $\boldsymbol{M}$ throughout. A magnet actually comprises a large number of elements ΔV, each of which has a volume $\delta x \cdot \delta y \cdot \delta z$ in Figure 1.3. Two adjacent elements may have different magnitudes for $\boldsymbol{M}$, and we shall consider only the z components of these shown in Figure 1.3, M_z and M'_z. For the element with M_z, Equation (1.7) gives

$$\sum \mu_{mz} = M_z \, \delta x \, \delta y \, \delta z \tag{1.9}$$

Since μ_{mz} is constant over an elemental volume, Equation (1.4) may be used to give

$$\sum \mu_{mz} = i \, \delta x \, \delta y \tag{1.10}$$

Considering that $\boldsymbol{M}$ was defined directly from $\boldsymbol{\mu}_m$, it is hardly surprising to find that $\boldsymbol{M}$ in an element is related to a circulating current i, as shown in Figure 1.3. Combining Equations (1.9) and (1.10), that relationship is

$$i = M_z \, \delta z \tag{1.11}$$

Similarly, in the neighboring element with M'_z there is

$$i' = M'_z \, \delta z \tag{1.12}$$

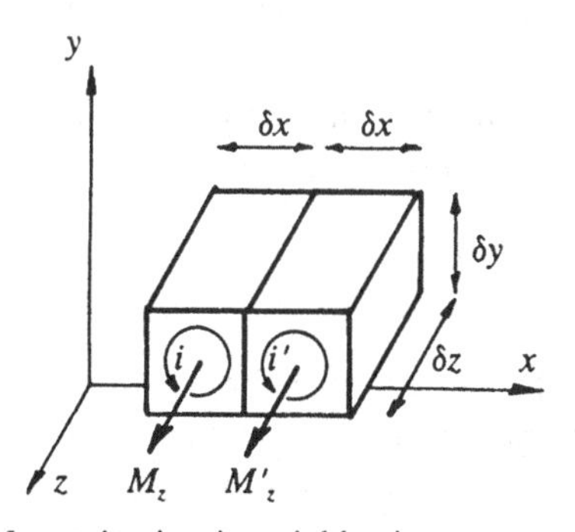

Figure 1.3. Magnetization in neighboring magnet elements.

The change from M_z to M'_z occurs over a distance δx, so the corresponding change in current may be expressed as

$$i' = i + \frac{\partial i}{\partial x}\delta x \qquad (1.13)$$

With the aid of Equation (1.11), this current flowing up the wall that divides the two elements may be written as

$$i - i' = -\frac{\partial M_z}{\partial x}\delta x\,\delta z \qquad (1.14)$$

This "wall" may be thought of as having an area $\delta x \cdot \delta z$ across which the current $i - i'$ flows in the y direction. This is a current density, given by

$$J_y = -\frac{\partial M_z}{\partial x} \qquad (1.15)$$

Now let us consider that $\boldsymbol{M}$ has a component in the x direction as well as that in the z direction. A change in M_x will also contribute to J_y (although a y component of $\boldsymbol{M}$ would not produce an equivalent current in its own direction). A similar derivation gives

$$J_y = +\frac{\partial M_x}{\partial z} \qquad (1.16)$$

The total y component of current density will be

$$J_y = \frac{\partial M_x}{\partial z} - \frac{\partial M_z}{\partial x} \qquad (1.17)$$

According to this expression, it is a change in the magnetization that is equivalent to the flow of current. J_y is just one of the three possible components of $\boldsymbol{J}_{\mathrm{m}}$ (J_x and J_z are the others), the current density which may be used to model $\boldsymbol{M}$. A full derivation of $\boldsymbol{J}_{\mathrm{m}}$ yields a vector equation, which separates these three components using unit vectors $\boldsymbol{i}$, $\boldsymbol{j}$ and $\boldsymbol{k}$ in the x, y and z directions respectively:

$$\boldsymbol{J}_m = \boldsymbol{i}\left(\frac{\partial M_z}{\partial y} - \frac{\partial M_y}{\partial z}\right) + \boldsymbol{j}\left(\frac{\partial M_x}{\partial z} - \frac{\partial M_z}{\partial x}\right) + \boldsymbol{k}\left(\frac{\partial M_y}{\partial x} - \frac{\partial M_x}{\partial y}\right) \qquad (1.18)$$

The $\boldsymbol{j}$ component in the y direction is the one that was derived in Equation (1.17). Fortunately, there is a shorthand notation for the operation that is performed upon a vector such as $\boldsymbol{M}$ in Equation (1.18), to yield another vector such as $\boldsymbol{J}_{\mathrm{m}}$. It is the *curl* operation, sometimes referred to as *rotation*, which is written mathematically as "$\nabla \times$" (Shercliff, 1977). Equation (1.18) is therefore written as

$$\boldsymbol{J}_{\mathrm{m}} = \nabla \times \boldsymbol{M} \tag{1.19}$$

As an example of this operation, imagine that there is a *rotation* of the $\boldsymbol{M}$ vector in the magnet, which causes the M_z component to change between adjacent elements, as was shown in Figure 1.3. The result is an equivalent current density, which we found by Equation (1.15) to be J_y, but which we now know to be just one component from the more general expressions (1.18) and (1.19).

If the magnet is *uniformly* magnetized, there will only be a change in $\boldsymbol{M}$ at the boundaries as shown in Figure 1.4, and there is only an effective rotation of $\boldsymbol{M}$ at the sides, which is where the equivalent current density $\boldsymbol{J}_{\mathrm{m}}$ will flow. A uniformly magnetized magnet is therefore exactly equivalent to a solenoid coil, as one might indeed expect.

Another well-known situation in electromagnetics to which the *curl* operation applies is the magnetic field $\boldsymbol{B}$ circulating (rotating) around a conductor carrying current – *Ampère's law* (Shercliff, 1977). $\boldsymbol{B}$ is related to the current density $\boldsymbol{J}$ through the constant μ_0 by

$$\mu_0 \boldsymbol{J} = \nabla \times \boldsymbol{B} \tag{1.20}$$

If $\boldsymbol{J}$ is only in the y direction with just a $\boldsymbol{j}$ component as shown in Figure 1.5, then J_y will have the same form as Equation (1.17):

$$\mu_0 J_y = \frac{\partial B_x}{\partial z} - \frac{\partial B_z}{\partial x} \tag{1.21}$$

As a *real* current density $\boldsymbol{J}$ causes a field $\boldsymbol{B}$ to circulate around it, then so too does an *equivalent* current density $\boldsymbol{J}_{\mathrm{m}}$ representing magnetization. When both conducting and magnetic media are present, the total flux

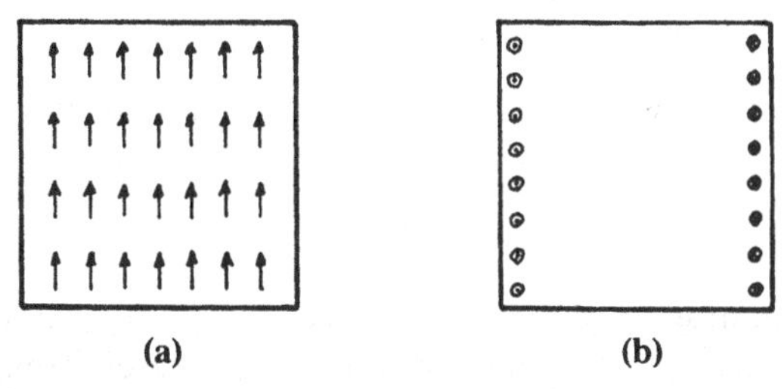

Figure 1.4. Uniformly magnetized magnet (*a*), which may be modeled by a current density over its boundary (*b*).

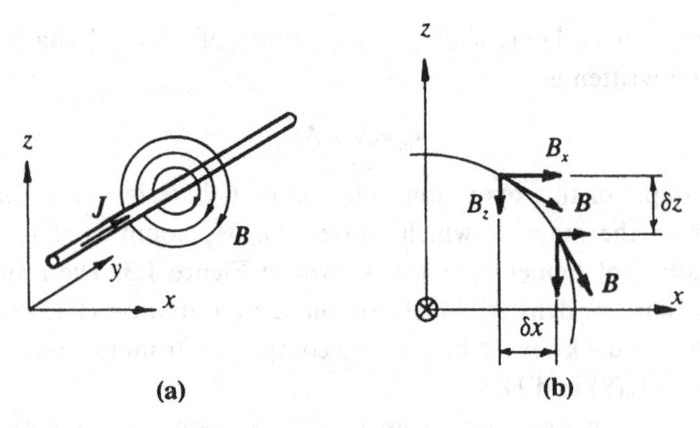

Figure 1.5. Circulation of flux density $\boldsymbol{B}$ around a conductor carrying current density $\boldsymbol{J}$.

density due to both sources will be

$$\mu_0(\boldsymbol{J}+\boldsymbol{J}_{\mathrm{m}})=\nabla\times\boldsymbol{B} \tag{1.22}$$

However, $\boldsymbol{J}_{\mathrm{m}}$ itself is caused by the material magnetization $\boldsymbol{M}$ according to Equation (1.19), so

$$\mu_0(\boldsymbol{J}+\nabla\times\boldsymbol{M})=\nabla\times\boldsymbol{B} \tag{1.23}$$

This expression may be rearranged to group the two magnetic parameters together as

$$\mu_0\boldsymbol{J}=\nabla\times(\boldsymbol{B}-\mu_0\boldsymbol{M}) \tag{1.24}$$

In the previous Section we showed that *spontaneous magnetization* could alone cause a field within the material given by $\boldsymbol{B}=\mu_0\boldsymbol{M}$, which is the condition in Equation (1.24) for *real* currents to be absent ($\boldsymbol{J}=0$). In reality, we will be interested in the *external* effects that a permanent magnet can produce for us, which will of necessity require $\boldsymbol{B}$ to be dissimilar to $\mu_0\boldsymbol{M}$. To this end, a new parameter called *magnetizing force* $\boldsymbol{H}$ is defined as

$$\mu_0\boldsymbol{H}=\boldsymbol{B}-\mu_0\boldsymbol{M} \tag{1.25}$$

The more common way to write this most fundamental relationship between the three macroscopic parameters of a magnet is

$$\boldsymbol{B}=\mu_0(\boldsymbol{H}+\boldsymbol{M}) \tag{1.26}$$

This allows Equation (1.24) to be rewritten as a more general version of Ampère's Law, one that allows for real currents and material magnetization:

$$\boldsymbol{J}=\nabla\times\boldsymbol{H} \tag{1.27}$$

In a magnetic material, an applied current (density $\boldsymbol{J}$) actually causes $\boldsymbol{H}$, rather than $\boldsymbol{B}$, though for a non-magnetic medium we find that Equation (1.27) reverts to Equation (1.20) by setting $\boldsymbol{M}=0$ and using $\boldsymbol{B}=\mu_0\boldsymbol{H}$. If a winding is placed around a permanent magnet and electrical current is applied, then the magnet becomes magnetized by virtue of $\boldsymbol{J}$ establishing $\boldsymbol{H}$ within the material. $\boldsymbol{H}$ in turn establishes $\boldsymbol{M}$ and hence $\boldsymbol{B}$, which explains it being called *magnetizing force*.

1.4 Magnetocrystalline anisotropy

Many permanent magnet materials are manufactured in a way that enhances their magnetic properties along a preferred axis, because in most applications we are only interested in field being produced in one particular direction through the magnet. The most fundamental way to attain this is if the crystal lattice structure of the material itself has preferred directions for the magnetic moments, which may then form the foundation for achieving net alignment in the magnet. Such an alignment of the magnetic dipole moments in the lattice is called *magnetocrystalline anisotropy*.

It is already seen in Equation (1.8) that the work done in rotating $\boldsymbol{\mu}_m$ in a material with magnetization $\boldsymbol{M}$ would have a minimum when $\boldsymbol{\mu}_m$ and $\boldsymbol{M}$ are aligned, and we can use this relationship as the basis for determining the preferred axes in a crystal lattice. It is helpful to rewrite Equation (1.8) using a trigonometric identity for $\cos\theta$ as

$$\boldsymbol{E}=-\mu_0\boldsymbol{\mu}_m M\left(1-2\sin^2\frac{\theta}{2}\right) \tag{1.28}$$

We can now define a *magnetocrystalline anisotropy energy* $\boldsymbol{E}_k$ as the *change* in energy that is required to rotate $\boldsymbol{\mu}_m$ from a preferred axis ($\theta=0$):

$$\boldsymbol{E}_k=2\mu_0\boldsymbol{\mu}_m M\left(\sin^2\frac{\theta}{2}\right) \tag{1.29}$$

In this expression, $\boldsymbol{E}_k$ is a maximum when $\theta=\pi$ (180°) and $\boldsymbol{\mu}_m$ is anti-parallel to $\boldsymbol{M}$ – an *unstable* condition.

The crystal lattices of real magnetic materials are much more complicated than this situation depicts, although Equation (1.29) can be easily modified to account for greater complexity. For example, iron, which is the principal element in many popular permanent magnets, has the body-centered *cubic* crystal lattice structure shown in Figure 1.6. There are now six equally preferred directions of magnetization: [0, 0, 1], [0, 1, 0], [1, 0, 0],

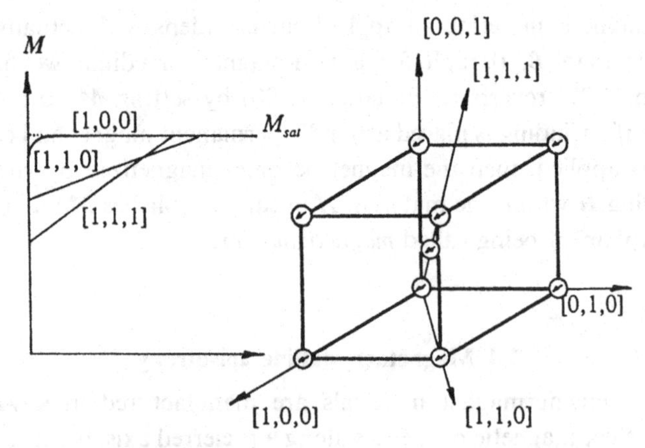

Figure 1.6. Body-centered cubic lattice structure of iron, and magnetization curves on various crystallographic axes.

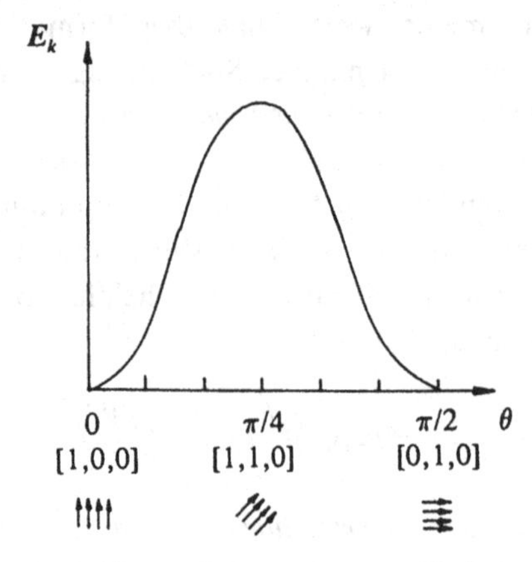

Figure 1.7. Magnetocrystalline anisotropy energy E_k in a cubic crystal lattice structure.

[0, 0, −1], [0, −1, 0] and [−1, 0, 0]. The *unstable* condition for μ_m is to lie in a plane between two of these at an angle of $\pi/4$ (45°) to each axis. It is only necessary to increase the periodicity in Equation (1.29) for it to apply to iron, as

$$E_k = 2\mu_0 \mu_m M \sin^2 2\theta \tag{1.30}$$

The plot of Equation (1.30) shown in Figure 1.7 demonstrates the stability of μ_m along two of the axes in one plane. In reality, rotation of μ_m may be considered in many other planes of the body-centered cube, such as its diagonals, which leads to Equation (1.30) being a special (and simplified) case of a more general relationship. Furthermore, other elements used in permanent magnet materials have more complex lattices; cobalt, for example, has a hexagonal structure. This model for iron, however, will suffice to develop an understanding of the basic characteristics of magnets: the magnetization curve, saturation, and coercive force.

A crystallographic constant for a magnetic element is generally defined as $K_1 = 8\mu_0\mu_m M$, and so Equation (1.30) is more commonly written as

$$E_k = \frac{K_1}{4}\sin^2 2\theta \tag{1.31}$$

K_1 is clearly a most fundamental parameter, which contains information on the suitability of an element for use in a permanent magnet. It is usually measured directly in an apparatus known as a *torque magnetometer*, shown in Figure 1.8. A small disk-shaped sample of the material is placed parallel to a uniform, constant, applied magnetic field, which is large enough to saturate the sample. The disk's circular shape ensures that the field within it will also be uniform. The sample is now rotated, but its magnetization remains in the direction of the saturating applied field. Because the material's magnetocrystalline anisotropy tries to keep its magnetization aligned in a preferred direction, rotation of the disk will create a restoring torque. The magnetometer measures this torque via the relative rotation

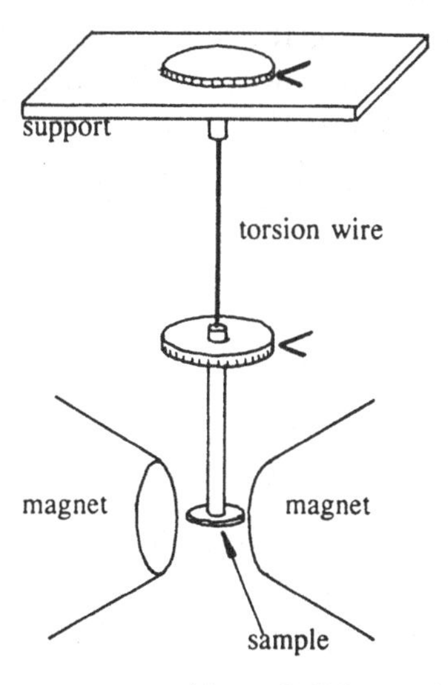

Figure 1.8. Torque magnetometer, with small disk sample of magnetic material.

of the opposite ends of a torsion wire to which the sample is attached, and the torque may be calculated using

$$\boldsymbol{T} = -\frac{\mathrm{d}E_k}{\mathrm{d}\theta} \tag{1.32}$$

Applied to Equation (1.31), the restoring torque due to magnetocrystalline anisotropy is

$$\begin{aligned} \boldsymbol{T} &= -\frac{K_1}{4} 4 \sin 2\theta \cos 2\theta \\ &= -\frac{K_1}{2} \sin 4\theta \end{aligned} \tag{1.33}$$

Applying a saturating field as in this experiment is useful for determining K_1, but in practical situations a magnetic material usually experiences external fields below this level. Consider our example of an elemental volume of iron, which is already spontaneously magnetized in its [1, 0, 0] preferred direction. This represents saturation of the material, with $\boldsymbol{M} = \boldsymbol{M}_{sat}$. Now an external magnetization force $\boldsymbol{H}$ is applied to the element at angle θ_0 to the [1, 0, 0] axis, as shown in Figure 1.9. If $\boldsymbol{H}$ were large enough to qualify as a *saturating* field, then all the magnetic moments would rotate to align with $\boldsymbol{H}$, as would $\boldsymbol{M}_{sat}$. However, for a general $\boldsymbol{H}$ lower than a saturating level, the moments will only rotate to an intermediate direction, which may be represented as $\boldsymbol{M}_{sat}$ turning through an angle θ ($<\theta_0$). The component of $\boldsymbol{M}_{sat}$ experienced in the direction of the applied field $\boldsymbol{H}$ will be

$$\boldsymbol{M} = \boldsymbol{M}_{sat} \cos(\theta_0 - \theta) \tag{1.34}$$

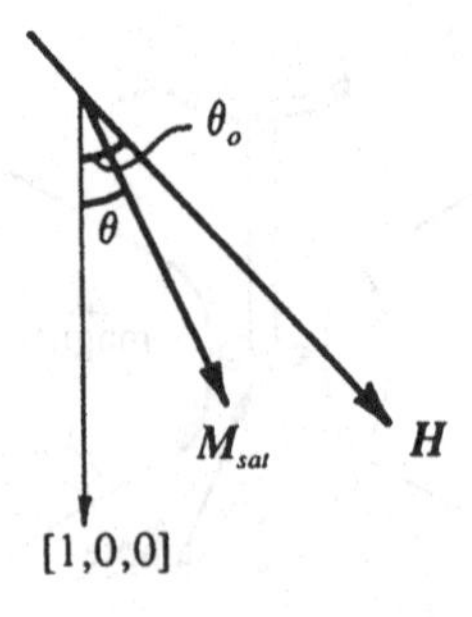

Figure 1.9. Rotation of $\boldsymbol{M}_{sat}$ from a preferred direction by an applied field $\boldsymbol{H}$.

In addition to the energy $\boldsymbol{E}_k$ associated with the crystal lattice structure, work is done by the applied field $\boldsymbol{H}$ in rotating the moments. This has already been derived as Equation (1.6) for a flux density $\boldsymbol{B}$ ($=\mu_0\boldsymbol{H}$) turning a single dipole, so the energy stored in the material due to an external field is

$$\boldsymbol{E}_f = -\mu_0\boldsymbol{\mu}_m H \cos\theta \tag{1.35}$$

It has also been shown that adjacent values and directions of $\boldsymbol{\mu}_m$ are identical in an elemental volume of the material, and may be summed via Equation (1.7) to yield $\boldsymbol{M}$; actually, this now corresponds to the definition of saturation magnetization $\boldsymbol{M}_{sat}$. According to Figure 1.9, $\boldsymbol{H}$ is now at angle $(\theta_0-\theta)$ to $\boldsymbol{M}_{sat}$, so the applied field energy per unit volume will be

$$\boldsymbol{E}_f = -\mu_0\boldsymbol{M}_{sat}H\cos(\theta_0-\theta) \tag{1.36}$$

The *total* energy stored will be the sum of $\boldsymbol{E}_k$ and $\boldsymbol{E}_f$:

$$\boldsymbol{E} = \frac{K_1}{4}\sin^2 2\theta - \mu_0\boldsymbol{M}_{sat}H\cos(\theta_0-\theta) \tag{1.37}$$

Optimum magnetic properties are obtained with $\boldsymbol{M}_{sat}$ aligned with a preferred axis, the [1, 0, 0] direction in this example. For the material to be a permanent magnet, this alignment should be able to withstand an external field of some magnitude in the opposite direction, in effect trying to *demagnetize* the material. This field H is contained within Equation (1.37), and so to evaluate this we may set $\theta_0=\pi$ (180°), and note that $\cos(\pi-\theta) = -\cos\theta$. Differentiating to find a minimum total energy yields

$$\frac{d\boldsymbol{E}}{d\theta} = \frac{K_1}{2}\sin 4\theta - \mu_0\boldsymbol{M}_{sat}H\sin\theta \tag{1.38}$$

Prior to application of $\boldsymbol{H}$, $\theta=0$ only, for which there is no solution to Equation (1.38) that will minimize this energy. However, since $\boldsymbol{H}$ is directly opposed to $\boldsymbol{M}_{sat}$, one is not looking for a gradual rotation through θ, but rather an applied field at which the magnetization will simply flip over into the opposite direction. This occurs when the *rate of change of energy* reverses, which is found from

$$\frac{d^2\boldsymbol{E}}{d\theta^2} = 2K_1\cos 4\theta - \mu_0\boldsymbol{M}_{sat}H\cos\theta \tag{1.39}$$

The value of H (applied at $\theta_0=\pi$) that causes $\boldsymbol{M}_{sat}$ to suddenly reverse is called the *intrinsic coercivity* of the material, H_{ci}. With $\theta=0$, Equation

(1.39) is zero when

$$H_{ci} = \frac{2K_1}{\mu_0 M_{sat}} \tag{1.40}$$

H_{ci} is one of the most important properties of a magnetic material, calculated only from its crystallographic constant and its saturation magnetization. Equation (1.40) provides a measure of the direct external demagnetizing force that the magnet can withstand. When the applied field reaches $-H_{ci}$ ($\theta_0 = \pi$), the magnetization flips over from $+M_{sat}$ to $-M_{sat}$; a further increase in the reverse field, however, will simply maintain the saturation level at $-M_{sat}$. If the applied field is reversed again, when it reaches $+H_{ci}$, the magnetization flips back from $-M_{sat}$ to $+M_{sat}$ and is maintained at that level for any further increase in the field. In this manner, we have just derived the *intrinsic magnetization characteristic* shown in Figure 1.10 for an elemental volume of a magnet, based upon its magnetocrystalline anisitropy.

The magnitude of H_{ci} is a measure of the permanent magnetism of a material, and it is also related to the saturation M_{sat}. The intrinsic characteristic of Figure 1.10 does not tell us how large H_{ci} *needs* to be, for a given value of M_{sat}, but a corresponding plot of flux density *B versus* magnetizing force *H* will provide information on the parameters most commonly used by magnet designers.

The *B versus H* loop in Figure 1.11 may be constructed from Figure 1.10 with the aid of Equation (1.26). Except at the two singularities, *M*

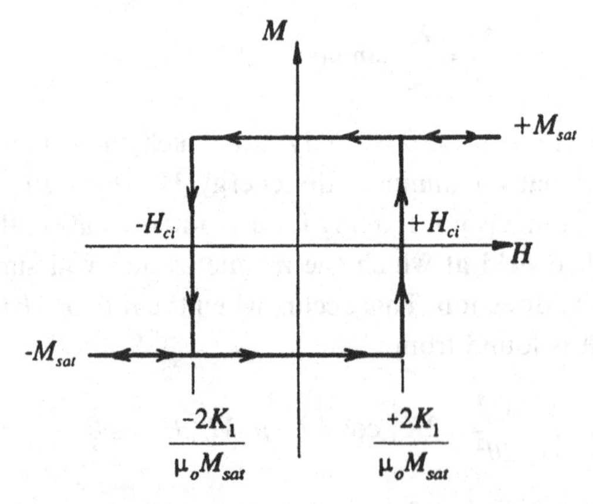

Figure 1.10. Intrinsic magnetization characteristic for an elemental volume of a magnet.

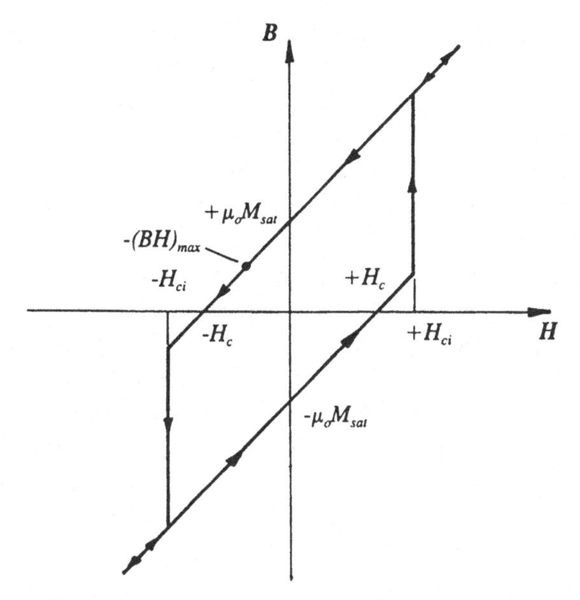

Figure 1.11. *B versus H* characteristic for a magnet.

always has a constant value of $+M_{sat}$ or $-M_{sat}$, so the slope of the B *versus* H curve is simply μ_0. The value of B when the magnetizing force is zero is called the *remanent flux density*, or more commonly the *remanence*, $B_r = \mu_0 M_{sat}$. The value of H that is required to reduce the flux density to zero is called the *coercive force* or *coercivity*, H_c. This is generally smaller than H_{ci}, provided that $|H_{ci}| > M_{sat}$, for which the magnitude of H_c is M_{sat}. This means that a smaller magnetizing force is required to remove B than is needed to reduce M to zero, so as a practical matter, the user of a permanent magnet would consider that value of H_c more important than that of H_{ci}. When $|H_{ci}| > M_{sat}$, the portion of the B *versus* H loop in the second quadrant of Figure 1.11 is entirely linear, since the *knee* which occurs at $-H_{ci}$ is moved out into the third quadrant. As we shall see, this is a highly desirable condition for application of a permanent magnet: the second quadrant alone describes the behavior of a magnet that is delivering flux $(+B)$ into an air gap $(-H)$.

1.5 Magnetic energy

As we have suggested, the B *versus* H loop is a measure of the energy that can be delivered by a permanent magnet, as indeed is the M *versus* H loop from which it is derived. While spontaneous magnetization alone will cause a field within the material, it will not allow the magnet to do work in the form of an external field. By its definition, it requires real currents to be absent, as provided by Equation (1.24) with $\boldsymbol{B}=\mu_0\boldsymbol{M}$, which condition also makes $\boldsymbol{H}=0$. Spontaneous magnetization therefore precludes operation on a *B versus H* or *M versus H* loop, only allowing alignment of the moments that is unassisted by an external field. This was illustrated by the magnetization curve in Figure 1.6 for the [1, 0, 0] preferred direction in iron. While magnetocrystalline anisotropy is a requirement for the material to be magnetic, the energy associated with an applied field is needed for the magnet to store energy or deliver work, as manifested by the magnetization loops of Figures 1.10 and 1.11.

The energy stored in a magnet due to an external field has already been derived as $\boldsymbol{E}_{\mathrm{f}}$ in Equation (1.35). The mechanism by which $\boldsymbol{H}$ transfers energy into and out of a magnet may be demonstrated using this. As before, the adjacent moments μ_{m} may be summed over a volume V of the material via Equation (1.7) to yield M, and one may consider also a preferred direction in which $\theta=0$, $\cos\theta=1$. The energy stored in volume V of the magnet may then be written as

$$E_{\mathrm{F}}=-\mu_0 VMH \tag{1.41}$$

Consider now the *change* in energy that is caused by a variation in $\boldsymbol{H}$:

$$\mathrm{d}E_{\mathrm{F}}=-\mu_0 VM\,\mathrm{d}H-\mu_0 VH\,\mathrm{d}M \tag{1.42}$$

The first term $-\mu_0 VM\,\mathrm{d}H$ is the work done by the applied field alone, experienced as the horizontal parts of the intrinsic magnetization characteristic in Figure 1.10. The second term $-\mu_0 VH\,\mathrm{d}M$ is the energy associated with a change in direction of the magnetic moments, these being the vertical transitions in Figure 1.10. Since the latter does not alter H, this difference represents kinetic energy within the material.

If the magnet is driven around a complete loop so that it returns to its original condition, such as point $\boldsymbol{a}$ in Figure 1.12, then $\oint \mathrm{d}E_{\mathrm{F}}=0$, so

$$-\mu_0 V\oint M\,\mathrm{d}H-\mu_0 V\oint H\,\mathrm{d}M=0 \tag{1.43}$$

This may be related to the *B versus H* loop by substituting Equation (1.26), then

$$-V\oint(B-\mu_0 H)\,\mathrm{d}H-V\oint H\,\mathrm{d}(B-\mu_0 H)=0 \tag{1.44}$$

$$\therefore\quad -V\left(\oint B\,\mathrm{d}H-\oint\mu_0 H\,\mathrm{d}H\right)-V\left(\oint H\,\mathrm{d}B-\oint\mu_0 H\,\mathrm{d}H\right)=0 \tag{1.45}$$

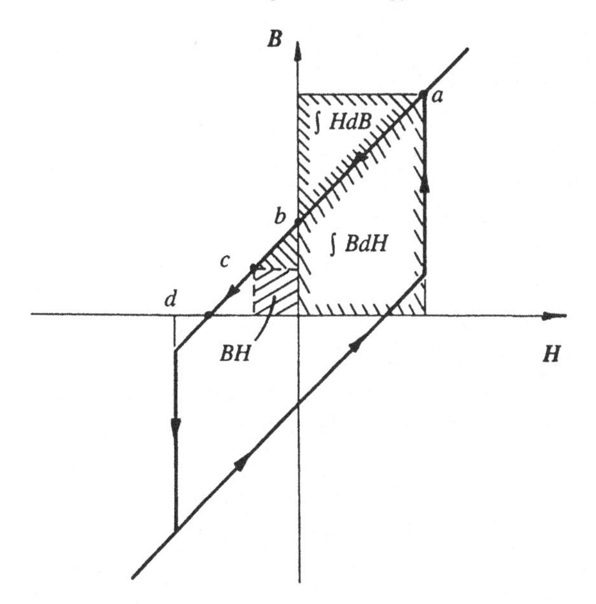

Figure 1.12. Energy changes around a *B versus H* loop.

Since H returns to its original value after a complete cycle, $\oint \mu_0 H \, dH = 0$, and Equation (1.45) reduces to

$$-\oint B \, dH - \oint H \, dB = 0 \tag{1.46}$$

The first term $-\oint B \, dH$ still gives the work done by the applied field, per unit volume of the material, and $-\oint H \, dB$ is the internal kinetic energy per unit volume.

To explain this mechanism further, consider a change from point ***a*** to point ***b*** in Figure 1.12, for which a summation of the areas yields

$$\int_b^a B \, dH + \int_b^a H \, dB = [BH]_b^a \tag{1.47}$$

$$\therefore \quad -\int_a^b B \, dH - \int_a^b H \, dB = -[BH]_a^b \tag{1.48}$$

Notice the similarities with the terms in Equation (1.46). This expression is still the total change in energy stored in a magnet due to an external field, and only differs from Equation (1.46) in that it allows the magnet to start and finish in different states. Actually, if either the applied field energy term $-\int B \, dH$ or the kinetic energy term $-\int H \, dB$ are integrated around the complete B *versus* H loop, each one represents the area enclosed by the loop, and the sum of both integrals is zero as required by Equation (1.46).

Each point on the B *versus* H loop therefore represents the total *potential* energy as $B \times H$. Although the magnet has been fully magnetized to saturation at point ***a***, the potential energy is reduced to zero at point ***b***; clearly the magnet can do no work when operating at its remanence. To deliver its stored energy, the magnet must move into the second quadrant of its B *versus* H loop, where H becomes negative but B is still positive; that is, the magnet experiences a *demagnetizing* force from the surrounding medium as it does work. Consider it changing from point ***b*** to point ***c*** in Figure 1.12. The total energy released increases with a change in the potential energy $-BH$. However, this change later decreases in magnitude as the magnet moves further into the second quadrant, as it approaches its coercivity at point ***d***, where again the potential energy is zero.

The product of $B \times H$ passes through a negative peak value in the second quadrant known as the *maximum energy product* $(BH)_{max}$. This quantity is the most common figure of merit for a permanent magnet. If point ***c*** corresponds to the maximum energy product, then the area in the rectangle shows that this operating point gives the maximum potential energy $-(BH)_{max}$ from the magnet. For the ideal material depicted in Figure 1.11, the second quadrant of the B *versus* H loop is linear, and $(BH)_{max}$ will occur at $B = \frac{1}{2}\mu_0 M_{sat}$, $-H = \frac{1}{2}M_{sat}$, with a value that is directly proportional to the saturation magnetization:

$$-(BH)_{max} = \mu_0 \left(\frac{M_{sat}}{2} \right)^2 \qquad (1.49)$$

This is the limiting value for the energy product, which is valid provided that $|H_{ci}| > \frac{1}{2}M_{sat}$. Equation (1.49) shows that $(BH)_{max}$ is directly related to the material's saturation magnetization, as was its intrinsic coercivity H_{ci}.

The second quadrant of the B *versus* H loop describes the condition of a magnet, which is delivering flux into an air gap, where the material experiences a demagnetizing force $-H$. It is helpful to consider the boundary between a magnet and air as shown in Figure 1.13, and the relationship between $\mathbf{B}$, $\mathbf{H}$ and $\mathbf{M}$ given in Equation (1.26). In air, where $\mathbf{M} = 0$, this simplifies to $\mathbf{B} = \mu_0 \mathbf{H}$, so the vector relationships on either side of the boundary are as illustrated in Figure 1.13. Across such a boundary, it is necessary for the magnetic flux to be continuous, because there can

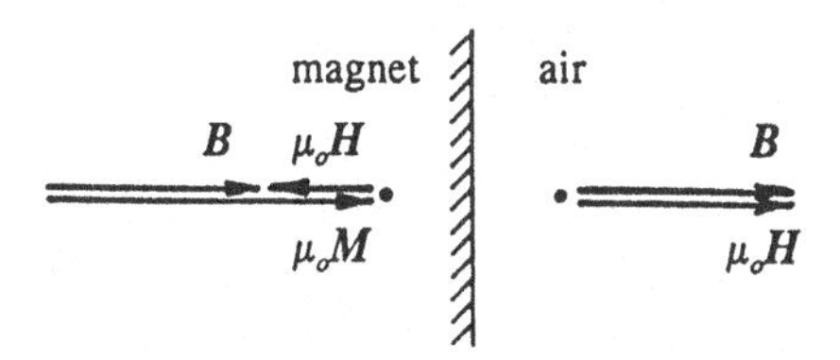

Figure 1.13. Field vectors either side of a magnet/air boundary.

be nothing contained within the boundary that could *create* flux (this would require the existence of a magnetic monopole!). Consequently, it is the flux density $\boldsymbol{B}$ that is continuous, as shown by its vectors in Figure 1.13. The magnetizing force $\boldsymbol{H}$ not only has an opposite direction either side of the boundary, but it will have a different magnitude (actually, as described later, it is not even necessary for $\boldsymbol{B}$, $\boldsymbol{H}$ and $\boldsymbol{M}$ to be aligned in the magnet because of their vector nature).

1.6 Shape anisotropy

To investigate magnetocrystalline anisotropy the torque magnetometer shown in Figure 1.8 was used, which tested a small disk-shaped sample of the material. It was mentioned that the sample's circular shape was important to establish a uniform field across it, although this is actually true for any particle with an ellipsoidal shape. It also required the applied field to be large enough to saturate the sample in the torque magnetometer experiment. In the general application of magnets, though, this is not the case, and as has just been described with reference to Figure 1.13, $\boldsymbol{H}$ within the material may differ substantially from the applied external field. Using the suffices i and o to represent the regions within and outside the material respectively, across the boundary in Figure 1.13 we may write

$$\boldsymbol{B}_{\mathrm{i}} = \boldsymbol{B}_{\mathrm{o}}$$

$$\therefore \quad \boldsymbol{H}_{\mathrm{i}} = \boldsymbol{H}_{\mathrm{o}} - \boldsymbol{M} \tag{1.50}$$

When dealing with the entire volume of a particle, the boundary is not as simple as this; consider, for example, the spherical particle shown in Figure 1.14. A uniform applied field produces a field within the sphere, which is also uniform because of its shape. Magnetic flux is clearly drawn into the magnetic medium from the surrounding air in the manner shown, so that the flux density $\boldsymbol{B}_{\mathrm{i}}$ within the sphere is greater than the applied

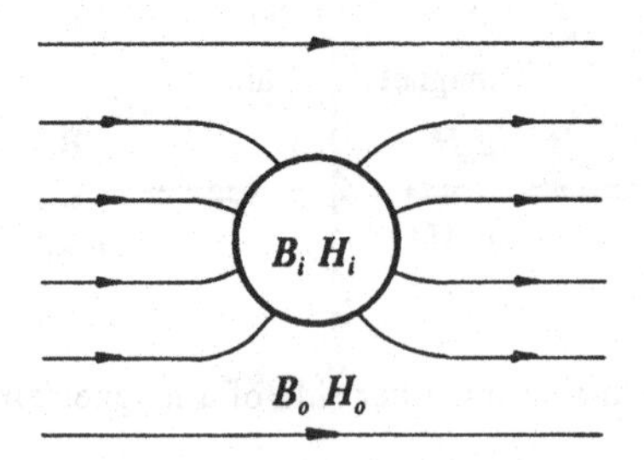

Figure 1.14. Uniform field in spherical magnetic particle.

flux density $\boldsymbol{B}_o$. Equation (1.50) is therefore no longer valid, and

$$\boldsymbol{B}_i > \boldsymbol{B}_o$$
$$\boldsymbol{H}_i > \boldsymbol{H}_o - \boldsymbol{M} \tag{1.51}$$

This relationship is generally written as

$$\boldsymbol{H}_i = \boldsymbol{H}_o - N\boldsymbol{M} \tag{1.52}$$

N is called the *demagnetizing factor* ($0 < N < 1$), and $N\boldsymbol{M}$ is a *self-demagnetizing field* of the magnetic particle. It is possible to solve magnetic field equations to determine the value of N for simple shapes that are either ellipsoids or close approximations to them. The simplest example is the sphere in Figure 1.14, for which it is calculated that $N = \frac{1}{3}$ (Bleaney and Bleaney, 1965). This approach permits our concept of spontaneous magnetization for when there is no external field $\boldsymbol{H}_o = 0$, and Equations (1.26) and (1.52) are combined to give

$$\boldsymbol{B}_i = \mu_0 \boldsymbol{M}(1 - N) \tag{1.53}$$

$N = 0$ is the limit where the self-demagnetizing field is absent, and spontaneous magnetization may most easily exist with $\boldsymbol{B}_i = \mu_0 \boldsymbol{M}$. The calculation yielding $N \approx 0$ is for a particle that is a very long, thin rod, magnetized along its major axis. The opposite poles are too far apart to create any significant self-demagnetizing field. $N = 1$ is the other limit, which may be calculated for a particle that is a very thin disk (or plate), magnetized normal to its plane. Opposite poles are now very close, establishing a strong self-demagnetizing field, which prevents spontaneous magnetization existing. Based upon this description, we would expect an ellipsoidal magnetic particle to exhibit anisotropy in the direction of its major axis. In fact, many popular types of permanent magnet are manufactured by processes that establish a net alignment of needle-shaped particles or platelets. These magnets base their properties on the *shape anisotropy* of their particles, rather than the magnetocrystalline anisotropy of the crystal lattice structure described earlier.

In our evaluation of magnetocrystalline anisotropy, the total energy stored in the material was the sum of the anisotropy energy $\boldsymbol{E}_k$ and the external field energy $\boldsymbol{E}_f$. Similarly, to evaluate a material's shape anisotropy, we must add the energy associated with the magnetic particles' shape, $\boldsymbol{E}_s$, to the energy stored by the external field $\boldsymbol{E}_f$. For an elemental volume ΔV of the material, $\boldsymbol{E}_f$ is still given by Equation (1.36). This expression may also be used to determine $\boldsymbol{E}_s$, if it is written to describe the rotation of moments from the major axis of an ellipsoid as shown in Figure 1.15, and if $\boldsymbol{H}$ means the internal field $\boldsymbol{H}_i$ used above. Combining Equations (1.34) and (1.36) gives

$$\boldsymbol{E}_s = -\mu_0 \boldsymbol{M}\boldsymbol{H}_i \tag{1.54}$$

Substituting Equation (1.52), and changing the applied field $\boldsymbol{H}_o$ to $\boldsymbol{H}$ to be consistent with Equation (1.36), we get

$$\boldsymbol{E}_s = -\mu_0 \boldsymbol{M}\boldsymbol{H} - \mu_0 N \boldsymbol{M}^2 \tag{1.55}$$

The demagnetizing factor is dependent upon the orientation of magnetization through the ellipsoid, and may be calculated as N_a and N_b along the major and minor axes respectively, shown in Figure 1.16. The magnetization vector $\boldsymbol{M}$ along each of these axes is found by resolving $\boldsymbol{M}_{sat}$ into its components $\boldsymbol{M}_{sat} \cos\theta$ and $\boldsymbol{M}_{sat} \sin\theta$, so $N\boldsymbol{M}^2$ in the second term of

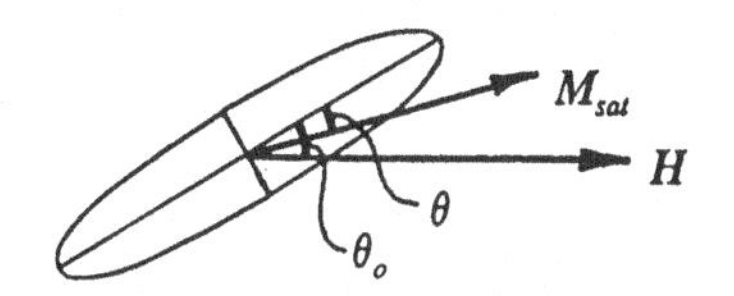

Figure 1.15. Rotation of $\boldsymbol{M}_{sat}$ from the major axis of an ellipsoid by a field $\boldsymbol{H}$.

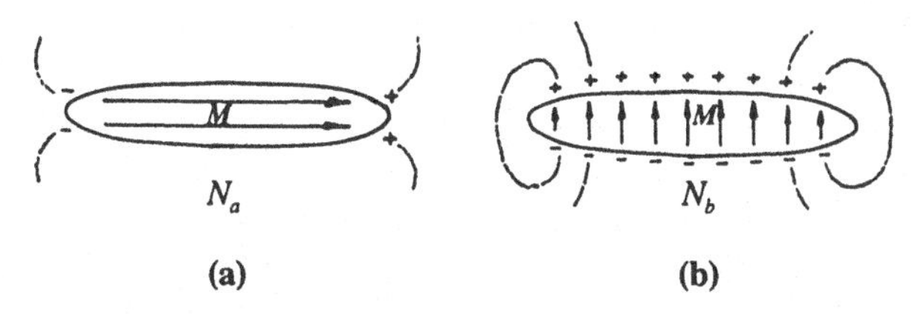

Figure 1.16. Demagnetizing factors in an ellipsoid, (*a*) N_a along major axis, (*b*) N_b along minor axis.

Equation (1.55) may be developed as

$$NM^2 = N_a(M_{sat}\cos\theta)^2 + N_b(M_{sat}\sin\theta)^2$$

$$= M_{sat}^2(N_a\cos^2\theta + N_b\sin^2\theta)$$

$$= \frac{M_{sat}^2}{2}[(N_b+N_a)-(N_b-N_a)\cos 2\theta] \tag{1.56}$$

Substituting this and Equation (1.34) into (1.55), the shape anisotropy energy may be written

$$E_s = -\mu_0 M_{sat} H\cos(\theta_0-\theta) + \frac{\mu_0 M_{sat}^2}{2}[(N_b+N_a)-(N_b-N_a)\cos 2\theta] \tag{1.57}$$

The total energy stored E will be the sum of E_s and E_f, Equations (1.36) and (1.57). To find the *intrinsic coercivity* of the material, we may investigate the applied field required to reverse the magnetization on the major axis of the ellipsoid, by setting H at $\theta_0=\pi$ (180°); with $\cos(\pi-\theta) = -\cos\theta$, E will be

$$E = +2\mu_0 M_{sat} H\cos\theta + \frac{\mu_0 M_{sat}^2}{2}[(N_b+N_a)-(N_b-N_a)\cos 2\theta] \tag{1.58}$$

Differentiating twice to find when the rate of change of energy reverses,

$$\frac{d^2E}{d\theta^2} = -2\mu_0 M_{sat} H\cos\theta + 2\mu_0 M_{sat}^2(N_b-N_a)\cos 2\theta \tag{1.59}$$

The intrinsic coercivity H_{ci} is the value of H (applied at $\theta_0=\pi$) that causes M_{sat} to suddenly reverse from $\theta=0$, for which Equation (1.59) is zero when

$$H_{ci} = M_{sat}(N_b-N_a) \tag{1.60}$$

This analysis shows that a magnetic particle may withstand an external field, which is applied so as to demagnetize it, depending upon the *shape* of that particle. The best result that can be achieved is along the major axis of a very long, thin platelet for which $N_a \approx 0$ and $N_b \approx 1$. Consequently, shape anisotropy will always yield $|H_{ci}| \leqslant M_{sat}$. When discussing magneto-crystalline anisotropy, it was noted that the condition for the *B versus H* loop to be linear throughout the second quadrant in Figure 1.11 was $|H_{ci}| > M_{sat}$, in which case the coercivity H_c will have magnitude M_{sat}.

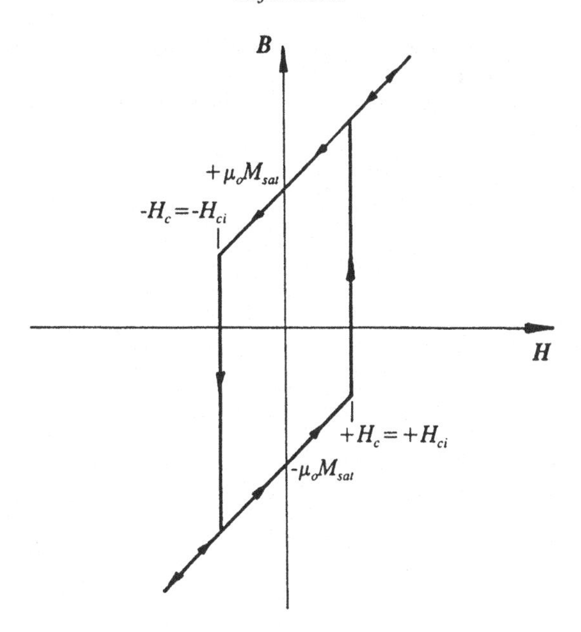

Figure 1.17. ***B*** *versus* ***H*** characteristic for a magnet with $|H_{ci}| < M_{sat}$.

Where shape anisotropy is the predominant mechanism, therefore, the *knee* that occurs at $-H_{ci}$ will occur in the second quadrant of the *B versus H* loop rather than the third, and $\pm H_c$ will be coincident with $\pm H_{ci}$. Since we can calculate the demagnetizing factors for ellipsoids, consider the case of a long, thin needle, for which $N_a = 0$ and $N_b = \frac{1}{2}$. These values give $|H_{ci}| = \frac{1}{2} M_{sat}$, and the *B versus H* loop will be of the form shown in Figure 1.17. While this has been a theoretical analysis of an ideal material, each type of anisotropy is seen to exhibit a particular shape of *B versus H* loop. The characteristics of real materials will not contain the abrupt changes shown in Figures 1.11 and 1.17. However, if a permanent magnet has an essentially linear second quadrant, then it is reasonable to assume that its performance is dominated by magnetocrystalline anisotropy. When discussing a magnet that is highly non-linear in this region, its low coercivity will have been caused by its dependence upon shape anisotropy.

References

Bleaney, B. I. and Bleaney, B. (1965). *Electricity and Magnetism*, 2nd edn. Oxford: The Clarendon Press.

Hadfield, D. (Ed). (1962). *Permanent Magnets and Magnetism*. New York: John Wiley & Sons.

McCaig, M. (1977). *Permanent Magnets in Theory and Practice.* New York: John Wiley & Sons.
Parker, R. J. (1990). *Advances in Permanent Magnetism.* New York: John Wiley & Sons.
Parker, R. J. and Studders, R. J. (1962). *Permanent Magnets and Their Application.* New York: John Wiley & Sons.
Shercliff, J. A. (1977). *Vector Fields.* Cambridge: Cambridge University Press.

2

Permanent magnet processes

2.1 Introduction

The shape of the B *versus* H characteristic of a material reveals whether its magnetism is based upon magnetocrystalline or shape anisotropy. In either case, the *ideal* characteristics described in Chapter 1 were founded on the concept of spontaneous magnetization, and this theory is certainly a good approximation for single crystals; measurements in the preferred [1, 0, 0] direction for iron shown in Figure 1.6 confirm this. However, practical materials do not follow this theoretical ideal, as shown for comparison by the initial magnetization curves for a real sample of iron in Figure 2.1. A measurable external applied field $\boldsymbol{H}$ is required to magnetize and saturate this material, so this sample does not exhibit spontaneous magnetization. The same is true for samples of nickel, cobalt and all alloys that are used to produce permanent magnets.

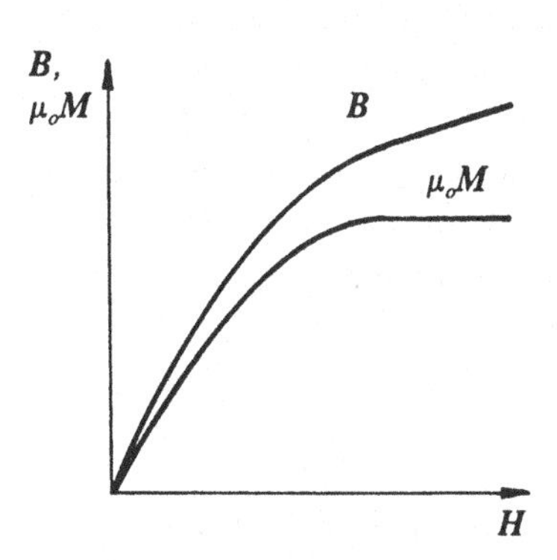

Figure 2.1. $\boldsymbol{B}$ *versus* $\boldsymbol{H}$ and $\boldsymbol{M}$ *versus* $\boldsymbol{H}$ initial magnetization curves for a bulk sample of iron.

While the magnetization curves for the single crystal in Figure 1.6 are dependent upon the crystallographic direction, the curve for a bulk sample of iron is not. A simple ferromagnetic material such as this is therefore isotropic with no preferred axis, and any enhanced properties in a specific direction will only be imparted to a magnet during its production process. Spontaneous magnetization still exists in the crystal lattice structure, so it must be explained how the theory of magnetocrystalline anisotropy must be modified to predict the actual characteristics of permanent magnets. To do this, the model of a magnetic material must first be enhanced.

2.2 Magnetic domains

The concept of the *magnetic domain* is of fundamental importance to ferromagnetism, because the size of individual domains will influence the magnetic properties of the material. Each domain contains some 10^{17} to 10^{21} atoms, it *is* spontaneously magnetized, and will therefore have a magnetic moment. The crystal lattice, however, comprises a large number of magnetic domains. In the absence of an applied field, there is a random orientation of these domains throughout the material, giving no net magnetization. As an external field $\boldsymbol{H}$ is generally applied, the domains' magnetic moments will gradually rotate to align with $\boldsymbol{H}$ and yield a magnetization $\boldsymbol{M}$ in the same direction. This traces an intrinsic curve such as that shown in Figure 2.1.

The domain concept therefore allows spontaneous alignment of magnetic dipole moments $\boldsymbol{\mu}_m$ within each domain and a magnetization $\boldsymbol{M}$ in each of these elemental volumes, as described in Chapter 1. However, this theory now goes further to explain the behavior of real magnets via gradual alignment of domain magnetizations to yield a net $\boldsymbol{M}$ for the bulk material. It is important to study the principal factors that influence domain size, not only because this adds to our understanding of the theory of magnets, but because it shows why it is desirable to obtain specific particle sizes for the production of high energy magnets.

The energy stored by a field in a domain is the same as in an elemental volume V of a magnet, derived as Equation (1.41), and the change in this energy was written as Equation (1.42):

$$E_F = -\mu_0 VMH \tag{1.41}$$

$$dE_F = -\mu_0 VM\,dH - \mu_0 VH\,dM \tag{1.42}$$

The first term was identified as the external field energy, while the second term was the internal energy associated with a change in direction of the magnetic moments. Consider a sample volume V, which is a single domain as in Figure 2.2(*a*), and is spontaneously magnetized. All the internal moments are aligned, so the change in stored energy dE_F will be due only to the first term of Equation (1.42), which is the external field connecting the free poles on one end of the sample to those on the opposite end. If the sample volume V is now divided into two separate domains as shown in Figure 2.2(*b*), which are spontaneously magnetized in opposite directions, it is clear that the external field is considerably reduced. While the external field term in Equation (1.42) is correspondingly reduced in magnitude, there will now be a contribution from the second term due to the reversal in direction of magnetization across the *wall* dividing the two domains.

A 180° reversal is most common in permanent magnet materials, but domains with other orientations may also exist. For example, Figure 1.6 showed that iron has six equally preferred directions, so a further possibility is to add closure domains at the ends of the sample in Figure 2.2(*c*). This arrangement appears to eliminate the external field, so the only stored energy in V is associated with the domain walls within the material. A more realistic arrangement for a larger sample is shown in Figure 2.2(*d*), in which the individual closure domains are smaller. The creation of magnetic domains reduces the external field energy, but it also establishes internal energy with the change in direction of magnetic moment across the walls. There is a stable state for the material when the sum of these energies is minimized.

The ideal magnetization characteristics in Figures 1.11 and 1.17 were based upon abrupt reversals of M_{sat} at the intrinsic coercivity, but the

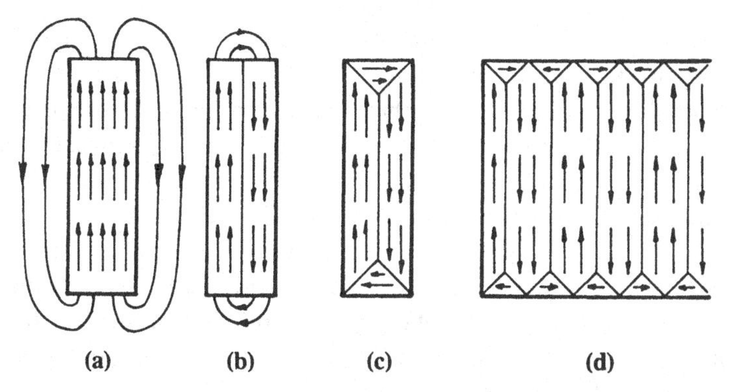

Figure 2.2. (*a*) A single domain sample of magnetic material, (*b*) two domains formed by a domain wall, and (*c*) and (*d*) the addition of closure domains to eliminate external fields.

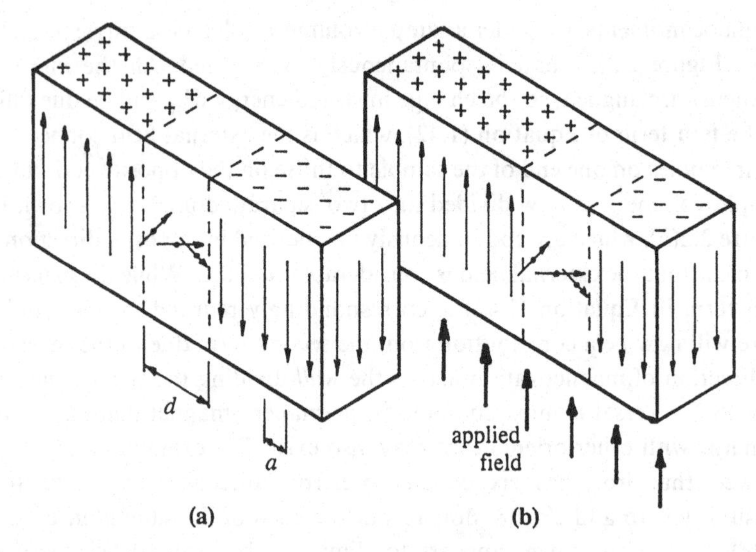

Figure 2.3. Domain wall of thickness d between two oppositely magnetized domains, (*a*) without and (*b*) with an externally applied field.

concept of magnetic domains allows these reversals to pre-exist in the material, within its domain walls. A reversal of magnetic moment cannot occur in adjacent atoms, however, and the transition takes place gradually as shown in Figure 2.3. A *wall* will have a finite thickness d through which there will be some 10^3 atoms. The rotation of moments is about an axis normal to the wall as shown here, which is required to leave no net moment normal to the wall (rotation about any axis parallel to the plane of the wall would yield a net moment and would require free poles to exist within the wall). This mechanism is confirmed by the predominance of 180° walls in magnetic materials. When an external field is applied to the material as in Figure 2.3(*b*), the moments in the walls are *gradually* rotated towards the direction of the field, so that the walls move. Domains having $\boldsymbol{M}$ parallel to $\boldsymbol{H}$ are enlarged at the expense of those with $\boldsymbol{M}$ anti-parallel to $\boldsymbol{H}$, which yields a net material magnetization.

In Chapter 1, we discussed the concept of an *exchange interaction* aligning the axes of neighboring magnetic dipoles, and wrote Equation (1.8) for the energy associated with aligning one dipole $\boldsymbol{\mu}_m$ with the net magnetization $\boldsymbol{M}$ from all its neighbors. Likewise, the work done in aligning two moments $\boldsymbol{\mu}_m$ may be written as

$$\boldsymbol{E}_e = -\mu_0 \mu_m^2 \cos\theta \tag{2.1}$$

The moments of two adjacent atoms in a wall will only contribute a small rotation $\Delta\theta$ and a small portion of the total exchange energy $\Delta \boldsymbol{E}_e$, so

$$\Delta \boldsymbol{E}_e = -\mu_0 \boldsymbol{\mu}_m^2 \cos(\Delta\theta) \tag{2.2}$$

Because $\Delta\theta$ is small, this may be written using an approximation for $\cos(\Delta\theta)$ as

$$\Delta \boldsymbol{E}_e = -\mu_0 \boldsymbol{\mu}_m^2 \left(1 - \frac{\Delta\theta^2}{2}\right) \tag{2.3}$$

The first term simply represents the normal exchange energy for parallel moments, as seen by setting $\theta = 0$ in Equation (2.1), so it is the second term that gives the increase in energy caused by transition through the wall. Consequently, the total increase in exchange energy across a 180° wall is

$$\boldsymbol{E}_e = \mu_0 \boldsymbol{\mu}_m^2 \sum_0^\pi \frac{\Delta\theta^2}{2} \tag{2.4}$$

Let there be n adjacent atoms through the thickness d of the wall, so if we assume that there is a steady rotation of the moments, then each neighboring pair will have the same $\Delta\theta$, and $n(\Delta\theta) = \pi$. This allows Equation (2.4) to be summed from 0 to n, giving

$$\boldsymbol{E}_e = \frac{\mu_0 \boldsymbol{\mu}_m^2 \pi^2}{2n} \tag{2.5}$$

The inter-atomic spacing is a, so $na = d$, but we may also divide by a^2 to find the total increase in exchange energy per unit area of the wall as

$$\boldsymbol{E}_e = \frac{\mu_0 \boldsymbol{\mu}_m^2 \pi^2}{2ad} \tag{2.6}$$

This expression shows that the thinner the wall is, the larger will be the energy increase $\boldsymbol{E}_e$, which explains why the 180° transition does not happen abruptly between two adjacent moments, but does occur gradually between a large number.

In addition to the exchange energy due to rotation of the moments through the domain wall, there will also be an energy associated with these moments not being aligned with the material magnetization $\boldsymbol{M}$. This *anisotropy energy* can be evaluated directly from Equation (1.8), with each moment $\boldsymbol{\mu}_m$ at angle θ to $\boldsymbol{M}$ contributing

$$\Delta \boldsymbol{E}_a = -\mu_0 \boldsymbol{\mu}_m \boldsymbol{M} \cos\theta \tag{2.7}$$

Using $n(\Delta\theta)=\pi$ and dividing by a^2, the total anisotropy energy per unit area across a 180° wall is found by integrating Equation (2.7) from $-\pi$ to 0, a transition from an anti-parallel domain to one that is aligned with $\boldsymbol{M}$:

$$\boldsymbol{E}_{\mathrm{a}} = -\frac{\mu_0\mu_{\mathrm{m}}Mn}{\pi a^2}\int_{-\pi}^{0} \cos\theta \,\mathrm{d}\theta \tag{2.8}$$

Substituting $na=d$, this yields

$$\boldsymbol{E}_{\mathrm{a}} = \frac{2\mu_0\mu_{\mathrm{m}}Md}{\pi a^3} \tag{2.9}$$

Wall thickness has opposite effects upon exchange energy $\boldsymbol{E}_{\mathrm{e}}$ and anisotropy energy $\boldsymbol{E}_{\mathrm{a}}$. $\boldsymbol{E}_{\mathrm{a}}$ increases with d, because the thicker the wall is, the more moments there will be that are not aligned with $\boldsymbol{M}$.

The existence of a wall between two domains with anti-parallel moments therefore creates a total increase in energy per unit area of the wall, which is the sum of Equations (2.6) and (2.9):

$$\boldsymbol{E}_{\mathrm{e}} + \boldsymbol{E}_{\mathrm{a}} = \frac{\mu_0\mu_{\mathrm{m}}^2\pi^2}{2ad} + \frac{2\mu_0\mu_{\mathrm{m}}Md}{\pi a^3} \tag{2.10}$$

A stable domain wall will have a thickness that gives the minimum of this total energy, which is found by differentiating Equation (2.10) with respect to d. The result is

$$d = \left(\frac{\mu_{\mathrm{m}}\pi^3 a^2}{4M}\right)^{1/2} \tag{2.11}$$

Substituting this back into Equation (2.10), the wall energy per unit area γ is found to be

$$\gamma = (\boldsymbol{E}_{\mathrm{e}} + \boldsymbol{E}_{\mathrm{a}})_{\mathrm{min}} = \frac{2\mu_0}{a^2}(\pi \boldsymbol{M}\mu^3)^{1/2} \tag{2.12}$$

It is worth noting that those materials with the thinnest domain walls have the highest anisotropy energies. For example, in soft ferromagnetic materials d is around 0.1 μm, whereas in rare earth permanent magnets, it is only about 0.002 μm.

To understand the practical significance of domain walls, it is helpful to consider the specific example of a uniformly magnetized spherical particle of radius r, which will contain a uniform field if it is a single

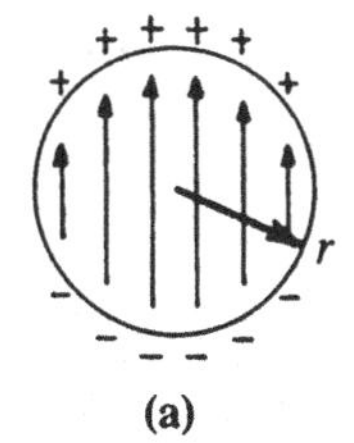

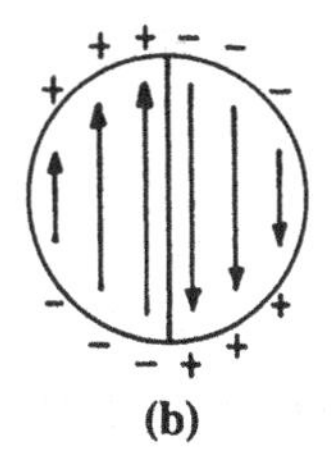

Figure 2.4. (*a*) A uniformly magnetized sphere of radius r, (b) divided into two hemispherical domains by a wall.

domain. Shown in Figure 2.4, we shall evaluate the condition for this to be divided into two oppositely magnetized hemispherical domains by a single wall. In Chapter 1, a *demagnetizing factor* was defined with reference to Equation (1.52), which may be calculated as $N=\frac{1}{3}$ for the sphere (Bleaney and Bleaney, 1965). When the particle is spontaneously magnetized and there is no external field, $\boldsymbol{H}_0=0$ and the internal field is found from Equation (1.52) to be $\boldsymbol{H}_i=-\frac{1}{3}\boldsymbol{M}$. This is used in Equation (1.42) to give the field energy stored within the sphere as

$$\boldsymbol{E}_F=\frac{\mu_0 V\boldsymbol{M}^2}{3} \tag{2.13}$$

The volume of a sphere is $\frac{4}{3}\pi r^3$, so

$$\boldsymbol{E}_F=\left(\frac{4\pi}{9}\right)\mu_0\boldsymbol{M}^2r^3 \tag{2.14}$$

When the wall is introduced as in Figure 2.4(*b*), the magnetization in each hemisphere will no longer be uniform, and it will become a much more complex matter to calculate the new field energy. However, it has already been explained with reference to Figure 2.2 that introduction of a wall would reduce the stored field energy (though this was compensated by the energy created in the wall). The decrease in $\boldsymbol{E}_F$ will be of the same form as Equation (2.14), and may be written using the constant δ as

$$\boldsymbol{E}_F=-\delta\mu_0\boldsymbol{M}^2r^3 \tag{2.15}$$

Creating a wall of area πr^2 will also introduce an energy of $\pi r^2\gamma$ (where γ was defined in Equation (2.12)), so the total energy change resulting from dividing the sphere into two domains is

$$\boldsymbol{E}=\pi\gamma r^2-\delta\mu_0\boldsymbol{M}^2r^3 \tag{2.16}$$

This shows that it is possible to define a critical radius r_c for which this energy is *zero*, as

$$r_c = \frac{\pi\gamma}{8\mu_0 \boldsymbol{M}^2} \tag{2.17}$$

If the radius of the particle is more than this critical value ($r > r_c$), then the energy change $\boldsymbol{E}$ is negative. Because the introduction of walls in such a particle will reduce the stored energy, a more stable state will be formed by creation of new domains. Conversely, if the particle's radius r is less than r_c, the introduction of new walls will increase the stored energy, so it will not be favorable for new domains to be formed. This information is used to prepare particles for use in permanent magnet materials. From the previous discussion and comparison between Figures 2.4(*a*) and (*b*), it is clearly preferable for there to be no domain wall, since the particle is then spontaneously magnetized and the maximum external field is derived from it.

A high coercivity magnet may then be produced, having strong magnetocrystalline anisotropy through alignment of the magnetic moments in the crystal lattice. Several types of permanent magnet are made by reducing the grain diameter so that only single domain particles can exist. Impurities will always exist in real magnet materials, but these are usually introduced deliberately to enhance the magnetic properties through an improvement in the coercivity.

2.3 Ceramic ferrite magnets

By far the most widely used type of permanent magnet material is the class known as *ferrites*. These are fine particle magnets made by powder metallurgical methods, so they are also known as *ceramic ferrites*. During their production the powder may be milled to particles that are approximately single domain size (about 1 μm diameter), which means that ceramic ferrites base their permanent magnetism on magnetocrystalline anisotropy. As was shown in Figure 1.11, this mechanism produces a magnet with an inherently high coercivity and an almost linear second quadrant *B versus H* characteristic, which is highly desirable for many major applications including electric motors. Their great popularity is mainly due, though, to the low cost and great abundance of their raw materials.

As their name implies, *ferrite* magnets are made using iron oxide (Fe_2O_3). Natural lodestone is a simple form of this magnet, whose formula

Fe_3O_4 may be regarded as $(FeO)(Fe_2O_3)$. In Figure 1.6, the body-centered cubic lattice structure of pure Fe, with spontaneous magnetization in one of its six preferred directions, was shown. Contrast this to Fe_3O_4, in which oxygen atoms are impurities that form a face-centered cubic structure as shown in Figure 2.5. Between these atoms are two different shapes of voids: tetrahedral and octahedral. Fe_3O_4 comprises one divalent and two trivalent ions to four O atoms, i.e. $Fe^{2+}Fe_2^{3+}O_4$, and of the 24 voids formed by 32 O atoms, eight are tetrahedral, which are occupied by Fe^{3+} ions, while 16 are octahedral, filled equally with Fe^{2+} and Fe^{3+} ions. The interaction between these ions depends upon their spacing, and because of the location of the voids, the magnetic moments of the Fe^{3+} ions in the tetrahedral voids are anti-parallel to those of the Fe^{3+} ions in the octahedral voids. It is therefore only the moments of the Fe^{2+} ions that have a net alignment, and they are solely responsible for the magnetization in Fe_3O_4.

We have used this example of lodestone to illustrate the use of impurities, in this case oxygen, in the formulation of a magnetic material. The structure of modern permanent magnets is even more complicated than this, and ferrites actually have a hexagonal crystal lattice structure (Heck, 1974), which exhibits a high degree of magnetic anisotropy. Their composition generally follows the formula $MFe_{12}O_{19}$, where M is usually either barium (Ba) or strontium (Sr). The function of the M ions is to bind the hexagonal lattice together.

This composition is formed by wet or dry mixing the correct proportions of iron oxide with a carbonate of Ba (or Sr), followed by a reaction at

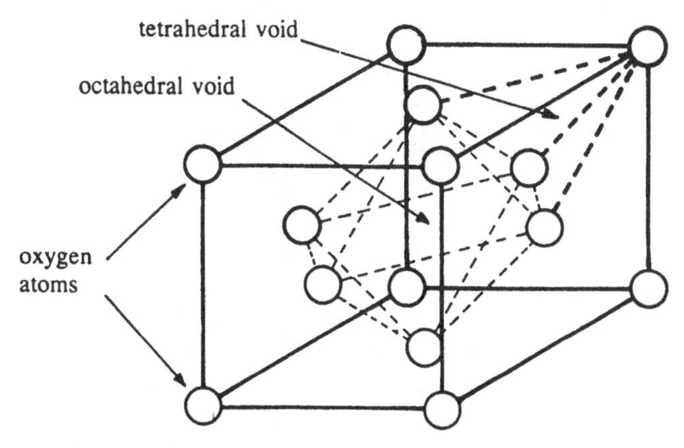

Figure 2.5. Face-centered cubic structure formed by oxygen atoms in Fe_3O_4.

between 1000 °C and 1350 °C, which may be considered to proceed in two stages:

$$\begin{aligned} BaCO_3 + Fe_2O_3 &\rightarrow BaOFe_2O_3 + CO_2 \\ BaOFe_2O_3 + 5\,Fe_2O_3 &\rightarrow BaO \cdot 6\,Fe_2O_3 \end{aligned} \tag{2.18}$$

The reacted mixture is next crushed and milled, and the powder is then wet or dry pressed in a die of the desired shape. If the object is to produce an anisotropic magnet with a preferred orientation, then it is important at this stage to attain milled particles of domain size. These are aligned during pressing by applying a magnetic field across the die cavity. Ferrite particles tend to develop as platelets with their planes perpendicular to the preferred hexagonal axis, so pressure will promote magnetic anisotropy in the pressing direction. Slightly better results are therefore achieved if the field is in the pressing direction, as shown in Figure 2.6. The compacted mixture is then sintered at a temperature in the range 1100–1300 °C, and the magnet is ground to its final dimensions. The production steps for ferrite magnets are summarized in Figure 2.7.

Ferrite permanent magnets are made in isotropic and anisotropic forms, and a wide range of properties are available. We have already mentioned many of the process variables, which include the size of the particles, the nature of impurities, the milling sequence, the duration and temperatures used in the reaction and sintering steps. The most straightforward process is sought for isotropic magnets, in which the milled powder is dried and pressed in a simple die that has no applied field. Anisotropic magnets can

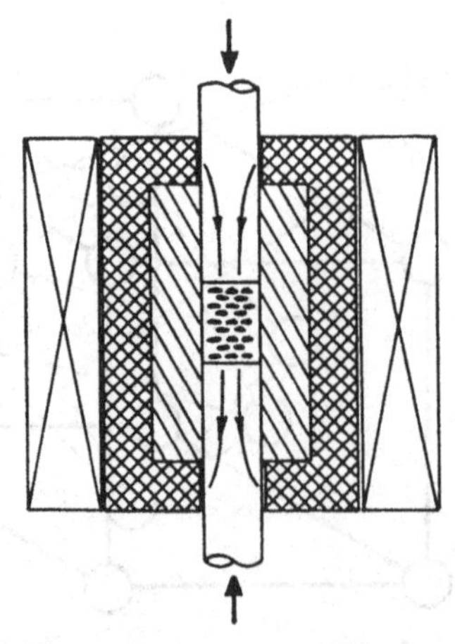

Figure 2.6. Ferrite particle orientation by magnetic aligning field in the press.

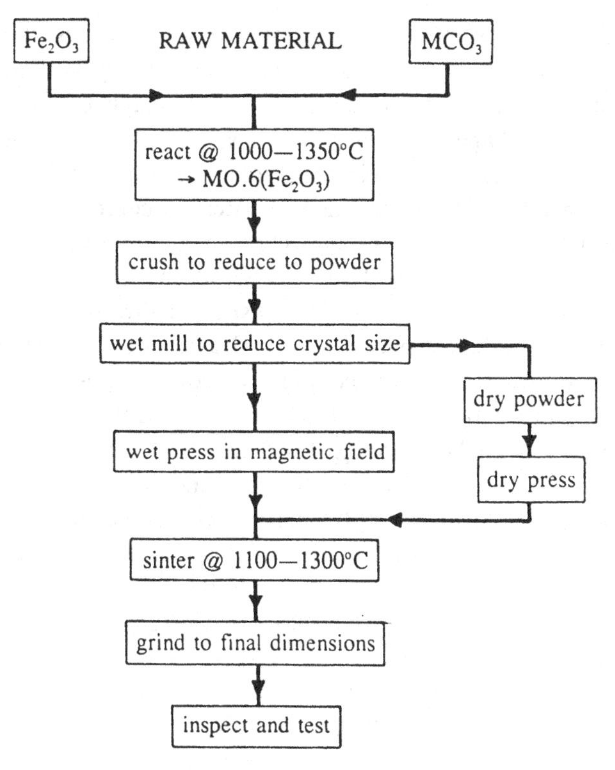

Figure 2.7. Production of ceramic ferrite magnets.

also be made with dried powder, but it is better to use a wet slurry in which water acts as a lubricant that promotes alignment of the powder as the magnetic field is applied in the die cavity. Shrinkage during sintering is significant, typically around 15%, and because the ceramic magnet is now very hard, grinding is really the only practical method of machining.

2.4 Alnico magnets

In Section 1.6, it was explained that an elongated particle would exhibit an enhanced coercivity along its major axis, a phenomenon known as *shape anisotropy*. This is the dominant mechanism in *alnico* permanent magnets, in which elongated magnetic particles are precipitated throughout the matrix of an **Al**–**Ni**–Fe–**Co** alloy. There are many variations on the constituents of this alloy, which produce different magnet characteristics, the earliest materials predating the inclusion of cobalt (which improves the anisotropy).

The most critical step in the processing of alnico magnets is a heat treatment of the alloy, which is controlled in such a way as to precipitate a dispersion of fine magnetic particles in a weakly magnetic matrix. These fine particles are shaped as long, thin rods, whose shape anisotropy is significantly stronger than their magnetocrystalline anisotropy; shape anisotropy, therefore, will determine the material's coercivity. During the cooling sequence, three different phases may crystallize out, known as the α_1, α_2 and γ phases. The objective is to obtain the α_1 phase, which is a weakly magnetic matrix of Al–Ni–Fe, dispersed throughout which are strongly magnetic α_2 phase particles of Co–Fe. Appearance of the γ phase will spoil this formation, but since it crystallizes at 1000–1100 °C, initial gentle cooling in air from around 1200 °C to around 900 °C can suppress this phase. Upon further cooling down to about 600 °C, the α phase decomposes into its α_1 and α_2 constituents, and the magnetic α_2 phase particles become elongated in a [1, 0, 0] crystallographic direction. In the final stage of the heat treatment, the alloy is tempered for 20–30 h at between 550 °C and 650 °C, which accentuates the difference in composition between the α_1 and α_2 phases, and allows the magnetic Co–Fe particles to grow even more elongated.

The structure of an unoriented Alnico 5 grade magnet is shown diagramatically in Figure 2.8(*a*), where the rod-shaped particles are approximately 40 nm × 8 nm × 8 nm, separated by about 20 nm. Since each particle has an improved coercivity along its major axis, the magnet itself will have enhanced magnetic properties in a unique direction if the particle axes are aligned. Such *anisotropy* is achieved by applying a magnetic field to orient the particles during their formation, i.e. while the α phase decomposes as it cools from 900 °C to 600 °C. This promotes the oriented structure shown in Figure 2.8(*b*).

By varying the heat treatment, a wide range of properties can be obtained for alnico magnets, but these also depend on whether the original

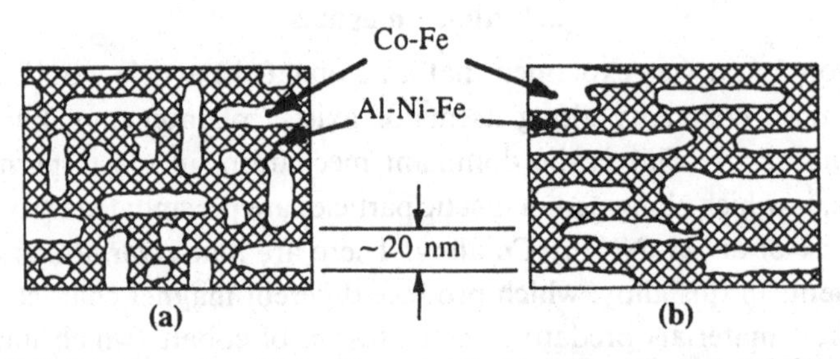

Figure 2.8. Structure of an Alnico 5 magnet that is (*a*) unoriented and (*b*) oriented.

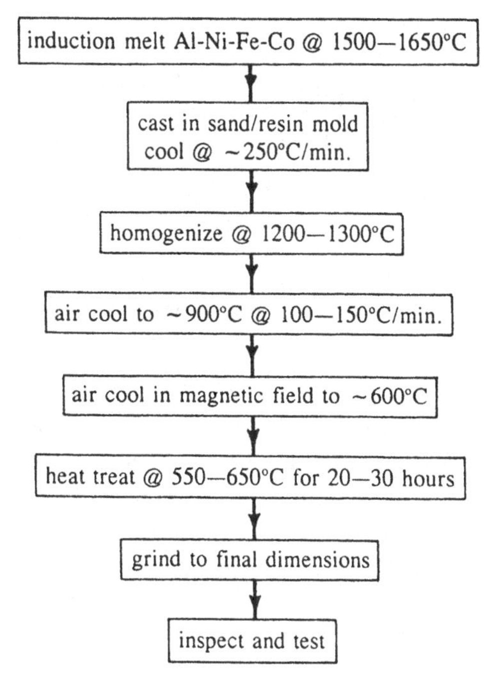

Figure 2.9. Production of cast Alnico 5 magnets.

alloy has been prepared by casting or by sintering. The complete production process for cast Alnico 5 is illustrated in Figure 2.9. The correct proportions of raw materials are melted in an induction furnace, and then cast into a sand and resin mold to the approximate final magnet shape. Casting must provide rapid cooling at around 250 °C/min to ensure that the components do not separate out in the mixture. An improvement in the directional properties of anisotropic alnico may be achieved by producing an elongated crystal structure in the casting operation itself. Alnico 5 DG is a *columnar* material in which directional grain (DG) growth is induced by withdrawing heat from the mold in the preferred direction only. This is typically done by casting into a special mold whose walls are heated by an exothermic reaction, but whose base forms a chill plate through which the alloy's heat is withdrawn (Figure 2.10). Because they are brittle, grinding is the most appropriate machining operation for cast alnico magnets. They typically receive a rough grinding prior to their heat treatment, and another after this to reach the final magnet dimensions.

Because casting is not a convenient process for small magnets, powder metallurgical techniques are employed as an alternative for production of

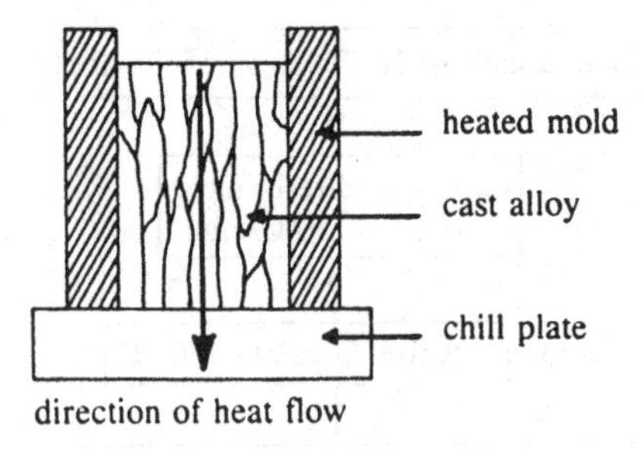

Figure 2.10. Casting of a columnar Alnico 5 DG magnet.

the alnico alloy. The constituent raw materials are in powder form, and because of its low melting point, Al is usually prepared as a prealloy with Fe, Ni or Co. The powders are mixed and then pressed in a die of the required shape, allowing for about 10% shrinkage during sintering. This sintering is performed at a temperature just below the melting point of the alloy. The final densities of sintered alnico magnets are slightly less than those made from cast alloys, a difference that is reflected in somewhat inferior magnetic properties. The cost of tooling for sintered materials is also higher, but they exhibit a more uniform fine grain structure, which makes them mechanically stronger than cast alnico magnets. The magnetic characteristics of alnicos are affected not only by their process variables, but also by minor changes in their composition. Cobalt itself was not included in the first Al–Ni–Fe materials, but it was soon discovered that Co raises both the saturation magnetization and the temperature stability of these magnets. It also delays the precipitation mechanism, which allows larger magnets of more uniform structure to be made. A few percent of copper is added to most Al–Ni–Fe–Co compositions; it enhances the magnetic properties, but must be used in conjunction with Co because it accelerates the precipitation process. It is also common to add around 5–8% titanium, which reduces the remanence somewhat, but also dramatically increases the coercivity. Alnico 8 is the most popular Al–Ni–Fe–Co–Ti magnet in current use.

2.5 Samarium–cobalt magnets

The desirability of a permanent magnet that has a high coercivity and an almost linear second quadrant *B versus H* characteristic, which is accomplished in a material having strong magnetocrystalline anisotropy, has already been mentioned. One method of assuring this is to utilize magnetic particles, which are of approximately single domain size, as with ferrite magnets. It has always been an important goal in magnet development to improve the properties while preserving these desirable features. When it appeared that no further significant improvements would be made to ferrite magnets, the search began in the 1960s for other materials that promised high uniaxial magnetocrystalline anisotropy, together with high saturation magnetization. It was known that the atoms of the rare earth elements exhibited incomplete electron shells, which induce ferromagnetism in much the same way as described for Fe_3O_4. It was also known that they tend to form intermetallic compounds with transition metals such as Fe, Ni or Co, and because of its hexagonal crystal structure, cobalt was at first expected to provide compounds with the highest crystal anisotropies.

The rare earth elements form a transition group with atomic numbers from 58 (Ce) to 71 (Lu), included within which may be Y (atomic number 39) and La (57), which have a similar behavior to rare earths in magnetic alloys. It is the light rare earth elements whose magnetic moments appear to combine most favorably with Co, particularly samarium (Sm). Processing of rare earth cobalt magnets allows various intermetallic compounds to be formed, including RCo_5, R_2Co_{17}, R_2Co_7, R_5Co_{19} or RCo_3 (where R represents one of the rare earth elements). RCo_5 was the first compound to be extensively investigated, and some early theoretical predictions indicated that very high values for *maximum energy product* might be expected from light rare earths such as La, Ce, Pr, Sm and also Y (Strnat, 1970). In the early years of development, however, it proved difficult to make satisfactory magnets from any of these compounds except for $SmCo_5$, which according to Table 2.1 quickly achieved a $(BH)_{max}$ close to its predicted limit.

Part of the difficulty in making these early magnets arose from the very high reactivity of the rare earths. The first process used to prepare $SmCo_5$ and similar powders was the *reduction/melt* procedure summarized in Figure 2.11 (Jones, 1987). This involved induction melting the constituents

Table 2.1. *Maximum energy product (kJ/m^3) for RCo_5, predicted and measured (Strnat, 1970).*

	YCo_5	$LaCo_5$	$CeCo_5$	$PrCo_5$	$SmCo_5$
Theoretical	225	165	118	288	187
Measured	12	46.5	—	34.4	160

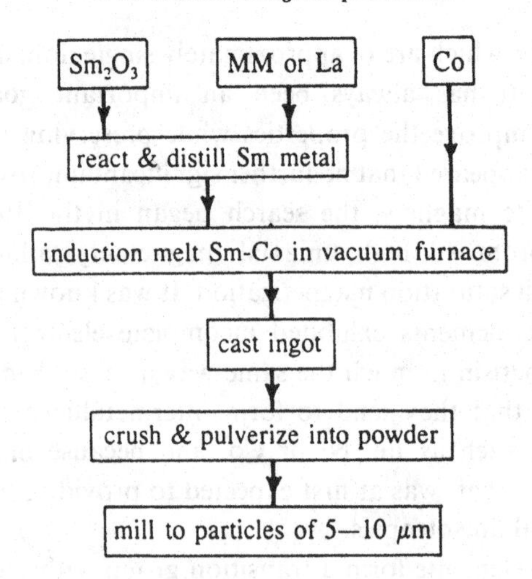

Figure 2.11. Reduction/melt process for samarium–cobalt powder.

together to form the alloy, but the crucible had to be made of high purity aluminum oxide and there also had to be protective atmosphere, such as a vacuum. After homogenization, the resulting cast alloy is very brittle, and it is readily crushed. Successive milling steps are then used to prepare a powder that may be formed into magnets using powder metallurgical techniques.

It has already been stated that the domain wall thickness d calculated by Equation (2.11) would be very small (around 0.002 μm) in rare earth magnets. This gives an extremely high wall energy per unit area, since $\gamma \propto d^{-2}$ (Equation (2.12)). Consequently, although d is very much smaller than in ferrites, for example, the critical radius for single domain particles r_c (Equation (2.17)) in rare earth–cobalts is not proportionately smaller than r_c for ferrites. Actually, for the variety of rare earth compounds in current use, r_c is typically in the range of 0.1–1.0 μm. According to the powder process described in Figure 2.11, it is *not* common practice to reduce rare earth alloy particles to single domain size, but rather to stop an order of magnitude larger at between 5 and 10 μm. With their grains milled within this size range, early $SmCo_5$ magnets were found to exhibit the highest coercivities corresponding to $(BH)_{max}$ values that were close to the theoretical limit given in Table 2.1. The simple *single domain* model used for earlier magnets was therefore not sufficient to fully describe the new rare earth types.

The explanation is that magnets like $SmCo_5$ allow a mechanism known as *nucleation* to occur. Since their grain diameter is somewhat greater than $2r_c$, it is energetically favorable for domain walls to be created, so domain wall motion will be relatively easy within each grain. This is shown by the initial magnetization curve of a previously unmagnetized "virgin" material, illustrated in Figure 2.12. While domain walls prevent *spontaneous magnetization* occurring, their ease of motion within the grains does allow saturation to be achieved with a reasonably small applied orienting field. High coercivity is representative of the magnet's ability to maintain its original magnetization when an external field is applied in the opposite direction, but once the *nucleus* of a reverse domain is formed, easy domain wall motion will allow the whole grain's magnetization to be reversed. High coercivity in a nucleation-type magnet must therefore be achieved at the grain boundaries, which are able to prevent domain wall motion progressing from grain to grain.

Real rare earth magnets are multi-phase systems with complex microstructures, and although the primary phase might be $SmCo_5$, the grain boundaries are usually the sites of deviations from this composition, which provide strong pinning of the domain walls. The alloy powders from the process of Figure 2.11 are formed into magnets using powder metallurgical techniques similar to those employed for ceramic ferrites. The powder is aligned with an applied orienting field while being compressed in a die or an isostatic press, then it is sintered at around 1100 °C to almost full densification. Composition deviations at the grain boundaries are enhanced by a post-sintering heat treatment, and annealing at about 900 °C appears to optimize domain wall pinning and maximize the coercivity of $SmCo_5$.

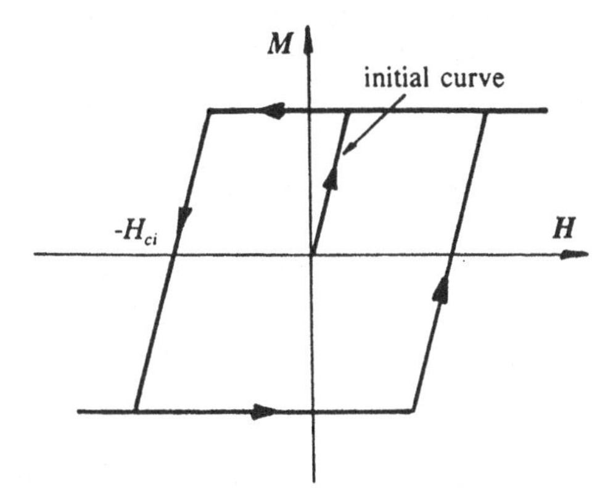

Figure 2.12. Intrinsic magnetization characteristic and the initial curve for a magnet exhibiting nucleation.

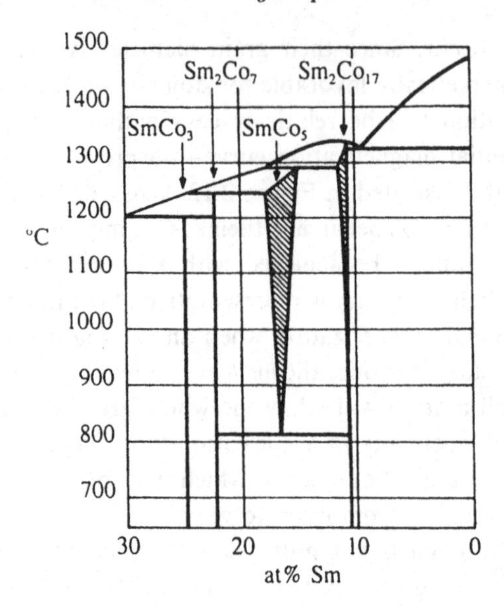

Figure 2.13. Samarium–cobalt phase diagram.

While various intermetallic compounds of samarium–cobalt are possible, its phase diagram shown in Figure 2.13 indicates a preference for Sm_2Co_{17} as well as $SmCo_5$. With less of the rare earth element, Sm_2Co_{17} has a higher saturation magnetization than $SmCo_5$, but in the early magnets this was not accompanied by high intrinsic coercivity. Various adjustments to the composition were made to improve the properties: iron is now added to increase M_{sat} and zirconium is used to raise H_{ci}. Values for $(BH)_{max}$ up to 240 kJ/m^3 were a significant improvement over those for $SmCo_5$ magnets. With the partial substitutions, the formulation is actually somewhere between $SmCo_5$ and Sm_2Co_{17}. Although still referred to as "2:17" magnets, the composition of modern materials is typically of the form $Sm(Co, Fe, Cu, Zr)_{7+x}$. Post-sintering heat treatments promote the difference between the Sm_2Co_{17} and $SmCo_5$ phases, and the development of a microstructure in which cells of Sm_2Co_{17} are separated by thin walls of $SmCo_5$. The addition of Cu is critical in enhancing a fine cell structure, as it appears to segregate only into the cell wall phase. These walls are only about 0.005 μm thick, which is comparable to the domain wall thickness.

Whereas "$SmCo_5$" was described as a nucleation-type magnet "Sm_2Co_{17}" does not rely upon the grain boundaries to block domain wall motion and create its coercivity. Rather, it is the $SmCo_5$ cell wall phase that pins the domain walls preventing their motion through the grains. *Pinning* replaces *nucleation* as the controlling process in these magnets, and grain size is of less importance (Livingston, 1986). The difference in microstructures of nucleation-type and pinning-type samarium–cobalt magnets is illustrated in Figure 2.14. It was the easy motion of domain walls through the grains that led to the low field required to initially magnetize a $SmCo_5$ magnet (Figure 2.12), but with effective pinning throughout the cells, a "virgin" Sm_2Co_{17} magnet requires a much greater initial orienting field and the typical magnetization characteristic is as shown in Figure 2.15.

While powder metallurgical techniques are the preferred method for making rare earth–cobalt magnets, an alternative approach to producing

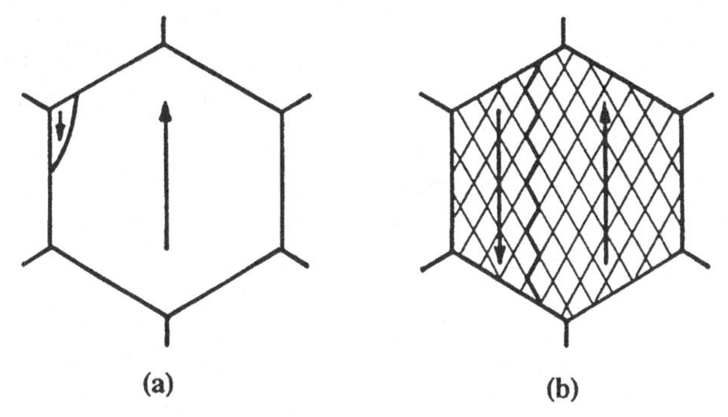

Figure 2.14. Illustration of samarium–cobalt magnet microstructures controlled by (*a*) nucleation, and (*b*) pinning in a fine cell precipitate.

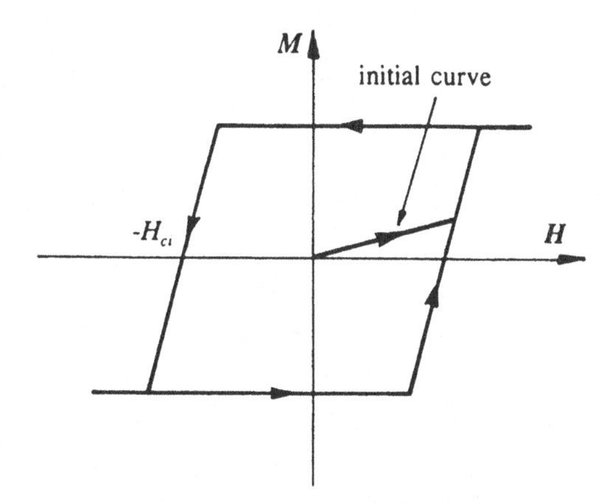

Figure 2.15. Intrinsic magnetization characteristic and the initial curve for a magnet exhibiting pinning.

the alloy powders has evolved. A more direct method than the *reduction/melt* procedure that was described with reference to Figure 2.11 is the *reduction/diffusion* process, which is summarized in Figure 2.16 (Jones, 1987). This is also a more economic process, because it starts with a rare earth oxide rather than a pure rare earth metal, this oxide now being *reduced* to metal with calcium, and *diffused* into cobalt powder. When applied to the production of $SmCo_5$ alloy powder, one chemical reaction that is used may be written as

$$10\,Co + Sm_2O_3 + 3\,Ca \rightarrow 2\,SmCo_5 + 3\,CaO \qquad (2.19)$$

After blending the constituents, the reaction is performed in an argon or hydrogen atmosphere at around 1150 °C, some further time being allowed for diffusion of Sm metal into the Co powder. Once this is completed, three steps are used to separate out CaO: a reaction with water, then gravimetric separation of the hydroxide, and finally an acid rinse to remove any traces of Ca. After drying, the $SmCo_5$ powder is ready for final milling to the desired particle size.

A variation on this reduction/diffusion process involves introducing some of the Co powder as cobalt oxide. The very exothermic nature of

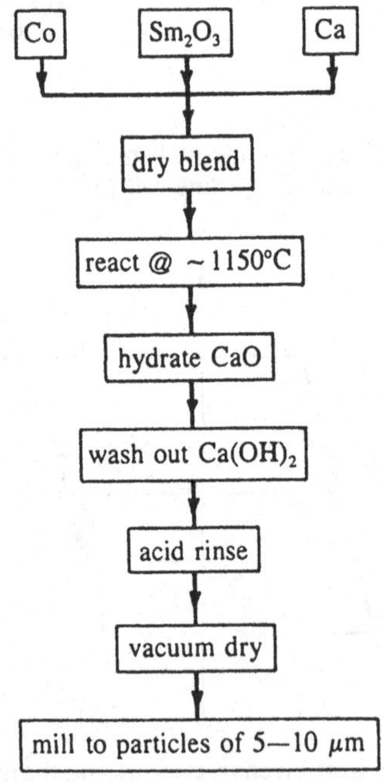

Figure 2.16. Reduction/diffusion process for samarium–cobalt powder.

the Ca reduction of cobalt oxide provides extra heat in the reaction, which is now performed in vacuum. A typical reaction for $SmCo_5$ in a *Co-reduction* process is

$$7\,Co + Co_3O_4 + Sm_2O_3 + 7\,Ca \rightarrow 2\,SmCo_5 + 7\,CaO \qquad (2.20)$$

but notice that much more Ca is now required, and more CaO must be separated out. Alloy preparation by various reduction/diffusion methods is now the most popular technique not only for samarium–cobalt powders but also for neodymium–iron–boron.

2.6 Neodymium–iron–boron magnets

After the successful development of $SmCo_5$ and then Sm_2Co_{17}, there was some concern that the cost and availability of the principal constituents might limit the commercial success of these magnets, with magnetic properties that were so superior to existing ferrite and alnico types. The first investigations involved using iron in place of cobalt with a variety of rare earth elements (R), but all the R_2Fe_{17} compounds have very low potential operating temperatures. We shall later define a fundamental characteristic of a magnetic material known as the *Curie temperature* (T_c), which is the temperature above which spontaneous magnetization will not exist. The practical operating temperature for a magnet is obviously well below T_c, and yet T_c itself is only around 125 °C for Sm_2Fe_{17} and around 60 °C for Nd_2Fe_{17}.

In the early 1980s, an important discovery was made, which modified R_2Fe_{17} to the ternary compound $R_2Fe_{14}B$, which has tetragonal crystal symmetry and a strong uniaxial magnetocrystalline anisotropy. The Curie temperatures for $R_2Fe_{14}B$ are some 200–300 °C higher than those of the corresponding R_2Fe_{17} compounds. Developments quickly focussed on $Nd_2Fe_{14}B$, which was predicted to offer the highest saturation magnetization, and whose T_c is slightly over 300 °C (Strnat, 1987). The most compelling attribute of this compound, however, is that neodymium is considerably more abundant than samarium, which, coupled with the use of iron as the transition metal, promised a significant saving in raw materials cost compared with samarium–cobalt magnets. Nevertheless, there is continued interest in using other more plentiful rare earths, such as cerium (Ce) or praseodymium (Pr), or even simply the natural mixture of the light rare earths which is called *mischmetal* (Y, La, Ce, Pr and Nd).

$Nd_2Fe_{14}B$ is the basic compound for the modern family of permanent magnets known as *neodymium–iron–boron*, but various partial substitutions and modifications are commonly made to adjust the magnetic properties to suit practical applications. Given that the rare earth–cobalt compounds have much higher T_c values, it is common to substitute a small portion of the Fe with Co to improve the properties at elevated temperatures, but this alone will also reduce the magnet's intrinsic coercivity. H_{ci} may itself be significantly improved by partial substitution of the Nd with a heavy rare earth element, the most popular of which is dysprosium (Dy). This is because $Dy_2Fe_{14}B$ has a higher anisotropy than $Nd_2Fe_{14}B$, but it also has an *anti*ferromagnetic coupling with Co, which reduces the saturation magnetization, and results in a reduction of the magnet's energy product, $(BH)_{max}$. Because many practical applications do require the properties of $Nd_2Fe_{14}B$ magnets to be more stable at higher temperatures, partial substitutions of Co and Dy are often performed together, these two elements somewhat compensating each other with respect to H_{ci}, while jointly improving the temperature stability. The variety of alloying elements and compositional variations that are available, coupled with the desire to improve temperature stability, has led to a wide range of "$Nd_2Fe_{14}B$" magnets being developed.

The performance benefits afforded by additional elements are somewhat negated by the extra complexity involved in producing an alloy other than the basic ternary compound. Reduction/diffusion methods for Nd–Fe–B powders are popular, one chemical reaction that is typically used being

$$57\ Fe + 8\ B + 10\ Fe_2O_3 + 7.5\ Nd_2O_3 + 52.5\ Ca \rightarrow Nd_{15}Fe_{77}B_8 + 52.5\ CaO \tag{2.21}$$

This is a *co-reduction* process, even though no cobalt is used in this example. The reaction is performed in vacuum, and the process steps closely follow those for samarium–cobalt powder as summarized in Figure 2.16. The proportions of Ca to rare earth oxide (7:1) and CaO to rare earth in the resulting alloy (3.5:1) are also the same as the previous example (Equation (2.20)). The milled Nd–Fe–B powder is aligned with an orienting field while being compressed in a die or isostatic press, then it is sintered to full densification, heat treated and ground to final dimensions.

Notice that the $Nd_{15}Fe_{77}B_8$ powder produced by the reaction of Equation (2.21) differs somewhat from the stoichiometric ratio for $Nd_2Fe_{14}B$; compositions that are used for practical sintered magnets are substantially Nd-rich and B-rich. Finished magnets usually contain non-magnetic secondary grain boundary phases into which the excess Nd and B diffuses, leaving the grain interiors as the highly magnetic $Nd_2Fe_{14}B$.

The grain boundaries therefore provide pinning of the domain walls, in the same manner as described for $SmCo_5$ magnets. Sintered $Nd_2Fe_{14}B$ is consequently a *nucleation*-type magnet, as shown by the initial curve of its intrinsic magnetization characteristic, which is again of the form shown in Figure 2.12. Its high saturation magnetization has allowed magnets of this type to be produced with $(BH)_{max}$ up to around 400 kJ/m^3.

A radically different alternative to the reduction/diffusion process for Nd–Fe–B powder is known as *melt spinning* or *rapid quenching*. An ingot of the required $Nd_2Fe_{14}B$ alloy is melted and forced under argon pressure through a small nozzle onto the surface of a water-cooled rotating metal wheel, producing a thin alloy ribbon as illustrated in Figure 2.17. Optimum magnetic properties, derived from the intrinsic coercivity that is achieved, depend upon the rate of quenching and hence upon the speed of the wheel (Croat, 1982). However, the window within which optimum results are obtained is very narrow, so it is usually necessary to anneal the quenched material to maximize the coercivity.

The ribbon produced is 1–3 mm wide and around 35 μm thick, and when milled to a powder, the particles are therefore shaped like platelets. Rapid quenching produces an extremely fine microstructure in the particles, containing individual $Nd_2Fe_{14}B$ grains of only around 0.030 μm diameter surrounded by secondary Nd-rich boundary layers, which are about 0.003 μm thick. The magnetic properties actually depend upon the quench rate via the grain size created, and this 0.030 μm diameter is an order of magnitude smaller than the single domain critical radius r_c, which is around 0.150 μm for $Nd_2Fe_{14}B$. Unlike powder produced by the reduction/diffusion process, rapidly quenched $Nd_2Fe_{14}B$ will *not* produce

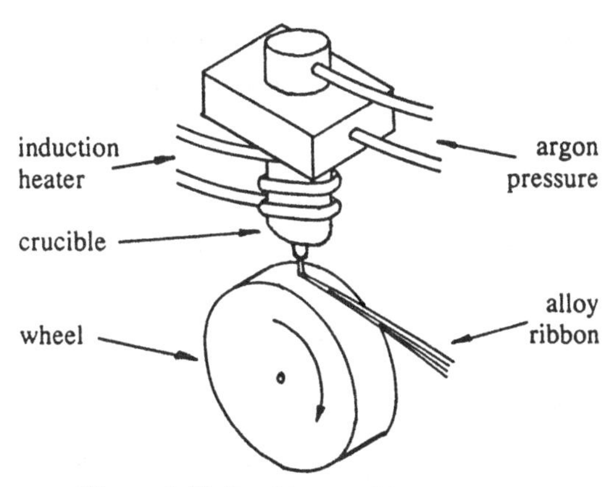

Figure 2.17. Rapid quenching process.

nucleation-type magnets, because this powder has a fine magnetocrystalline grain structure, which conforms to the *single domain* model described earlier.

In the manner of Figure 2.14, the microstructures of the two types of Nd–Fe–B magnet are illustrated in Figure 2.18. It is not practical to mill the rapid quenched particles to under 0.1 μm, which is required for them individually to be single domain grains. Rather, the random orientation of magnetization in the grains shown in Figure 2.18(*b*) makes these particles *isotropic*. The grains' magnetocrystalline anisotropy requires that an external field approximately equal to H_{ci} be applied before their magnetization vectors rotate into alignment, so, as shown in Figure 2.19, initial magnetization of rapidly quenched $Nd_2Fe_{14}B$ requires a much greater field than do nucleation-type magnets.

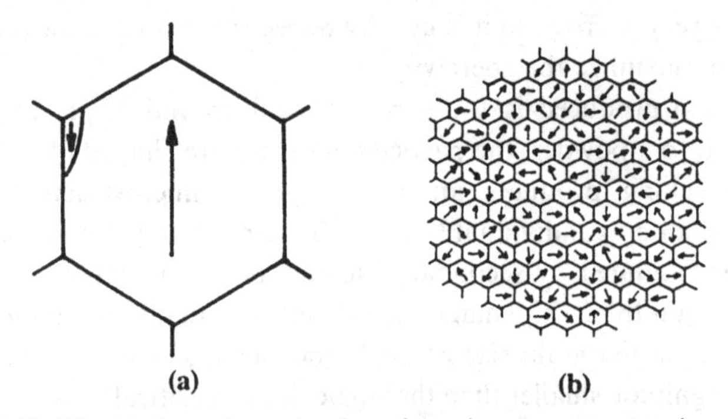

Figure 2.18. Illustration of neodymium–iron–boron magnet microstructures, which are (*a*) nucleation-type and (*b*) isotropic magnetocrystalline.

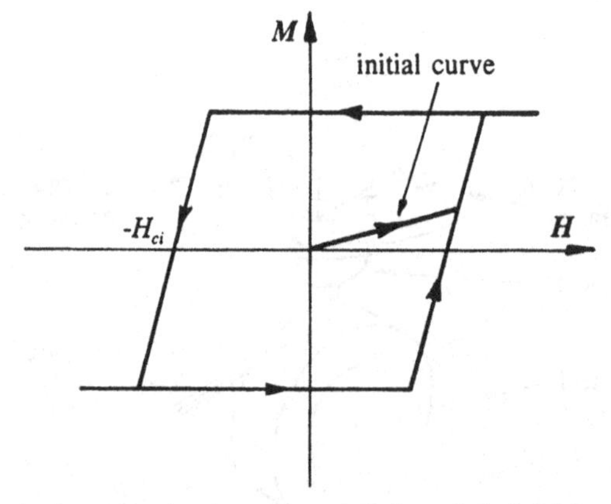

Figure 2.19. Intrinsic magnetization characteristic and the initial curve for a rapidly quenched $Nd_2Fe_{14}B$ magnet.

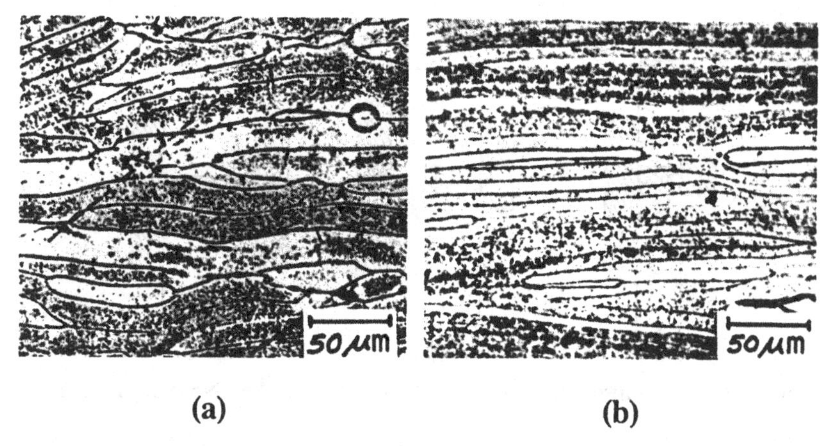

Figure 2.20. Compacts of rapidly quenched $Nd_2Fe_{14}B$, which are (*a*) hot pressed and then (*b*) die-upset. (Courtesy of Delco Remy, Division of General Motors.)

Since the rapidly quenched powder is not to be milled to single domain size, the actual size is chosen for convenience in the consolidation process. Platelets with long dimensions around 200 μm are typical, relative to their 35 μm thickness. Plastic flow of this material is achieved after only a few minutes at about 725 °C, under which conditions a *hot-pressed* magnet becomes fully dense with its platelet-shaped particles stacked as shown in Figure 2.20(*a*). This is a simple process since orientation of the isotropic particles by an applied field would give no benefit to the magnet properties. Without a preferred direction of magnetization, these essentially isotropic $Nd_2Fe_{14}B$ magnets can only achieve a $(BH)_{max}$ up to about 110 kJ/m^3.

A high degree of anisotropy can be imparted to these hot-pressed magnets by subjecting them to a technique known as *die-upsetting* (Lee, Brewer and Schaffel, 1985). The hot-pressed $Nd_2Fe_{14}B$ magnet is transferred into another die as illustrated in Figure 2.21, in which it is subjected to a further hot plastic deformation that reduces its height by around 50%. There is a lateral plastic flow in the die, transverse to the pressing direction, which significantly flattens the platelets as shown in Figure 2.20(*b*). The torque resulting from the internal shear stresses in the magnet causes the tetragonal lattice in the grains to rotate and slide such that the preferred magnetic axes in the lattice are perpendicular to the flow of material, parallel to the pressing direction. The alignment mechanism in die-upset rapidly quenched $Nd_2Fe_{14}B$ is purely crystallographic, and no external applied field is required to achieve strong anisotropy. With the grains' magnetization vectors aligned, the die-upset

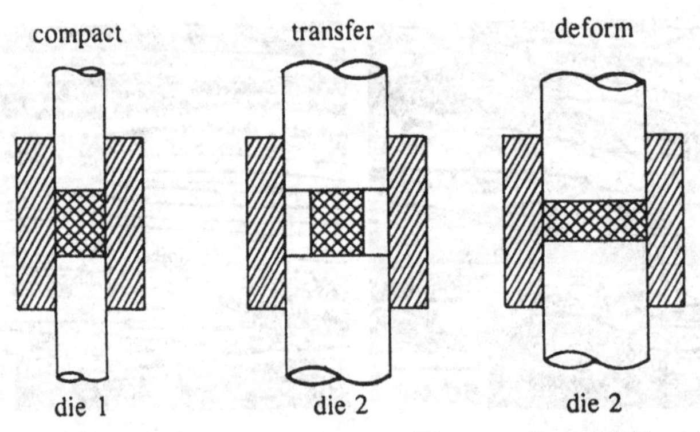

Figure 2.21. Die-upsetting process for a rapidly quenched $Nd_2Fe_{14}B$ magnet.

magnet will have a much higher saturation magnetization than the hot-pressed version, leading to a correspondingly higher $(BH)_{max}$ of up to around 320 kJ/m^3. This is comparable to the energy products achieved in commercial sintered nucleation-type $Nd_2Fe_{14}B$, except that the die-upset magnets have a much reduced intrinsic coercivity, H_{ci}. This is due to the very flat shape into which the platelets are formed; with reference to Equation (1.60), it has been shown that it is most difficult to demagnetize a platelet along its long axis, and conversely it is much easier to demagnetize it across its thickness.

One of the most troublesome aspects of the production of $Nd_2Fe_{14}B$ magnets is milling of the powder, which in the reduction/melt and reduction/diffusion processes of Figures 2.11 and 2.16 is performed until particles of 5–10 μm size are achieved. It is during this operation that oxygen is most likely to be picked up, the minimization of which greatly improves the stability of the powder. $Nd_{15}Fe_{77}B_8$ is known to absorb hydrogen very readily, particularly into the Nd-rich boundary phase, the effect of which is to turn the material into a very brittle powder, that can be milled to the required size much more efficiently. It is now common to include exposure to hydrogen in the production of $Nd_{15}Fe_{77}B_8$ powder, a process that is known as *hydrogen decrepitation* (HD).

An important modification to the HD process was reported in 1990 (Harris and McGuinness, 1990; Takeshita and Nakayama, 1990). By allowing decrepitation to occur in the hydrogen atomsphere at an *elevated* temperature of around 750 °C, a further reaction known as disproportionation takes place, in which the $Nd_2Fe_{14}B$ grain interiors become a very fine mixture of $NdH_{2.2}$, Fe and Fe_2B according to

$$Nd_2Fe_{14}B + 2.2\,H_2 \rightarrow 2\,NdH_{2.2} + 12\,Fe + Fe_2B \tag{2.22}$$

Next, the material is held under vacuum such that its hydrogen content is expelled, or *desorbed*. This causes the mixture to become thermodynamically unstable, and the remaining constituents of the grain interiors recombine into $Nd_2Fe_{14}B$. This procedure is known as *hydrogenation, disproportionation, desorption and recombination* (HDDR). The complete HDDR process effectively converts $Nd_{15}Fe_{77}B_8$ from a coarse grain structure into a material with an ultrafine structure having a grain size of around 0.3 μm. Furthermore, the new material is far more amenable to milling into a fine powder, comparable to its grain size.

The HDDR process produces a grain size comparable to the single domain critical radius for $Nd_2Fe_{14}B$ (0.15 μm), and like the powder produced by rapid quenching, the fine magnetocrystalline grain structure conforms most closely to the *single domain* model. Unlike rapid quenched powder, however, HDDR particles can be milled to around single domain size, and need not exhibit *isotropic* behavior as was illustrated in Figure 2.18(*b*). Partial substitutions for the Fe, such as Co and Zr, have been found to significantly enhance the anisotropy of the particles (Takeshita and Nakayama, 1992). Hot-pressing is the preferred route to producing fully dense magnets, since the temperature for plastic flow (around 725 °C) is comparable to that at which the optimum coercivity is achieved in HDDR powder. Anisotropic magnets can be made by two routes. In the same way that fully dense magnets are made from rapidly quenched $Nd_2Fe_{14}B$, isotropic HDDR powder can be hot-pressed into an isotropic magnet, and then subjected to hot plastic deformation in a die-upset process to achieve anisotropy. Alternatively, *anisotropic* HDDR powder can be hot-pressed directly into a fully dense magnet.

2.7 Bonded magnets

Many of the fully dense permanent magnet materials, especially those that are sintered, are very hard and brittle, and machining them to their final shape is often tedious. This led to an interest in *bonded* magnets, which are made by consolidating a magnet powder with a polymer matrix. While machining is easy, the production processes also frequently allow parts to be made directly to their final dimensions. Thermosetting binders, such as epoxy resin, are employed for use in compression-molded magnets, thermoplastic binders like nylon for injection-molded magnets, and elastomers such as rubber are used for extruded magnets.

The major drawback to bonded magnets is the reduction in their magnetic properties, relative to those that are 100% dense with magnetic material. Taken to the limit, a spherical magnet particle, which is effectively isolated from its neighbors in a polymer matrix, will behave as was shown in Figure 1.14. The demagnetizing factor in a sphere is $N=\frac{1}{3}$, so according to Equation (1.53) the flux density *in the particle* will only be two-thirds of its saturation level. For an ideal magnet with a linear second quadrant of its B *versus* H characteristic, Equation (1.49) shows that the corresponding $(BH)_{max}$ will be only $(\frac{2}{3})^2$ of the fully dense value. Clearly actual magnet particles are *not* isolated as this assumes, but then neither will the field in the particles represent the accumulated field from the bulk material.

Although without a theoretical derivation, an empirical relationship, which approximates to practical experience, states that the *remanence*, B_r, of a bonded magnet is reduced from that of its fully dense counterpart by the volume fraction, v, of the magnetic particles (Hopstock, 1987). A linear demagnetization curve again yields the greatest value for the energy product, and $(BH)_{max}$ would be correspondingly reduced by v^2. To optimize the magnetic properties of any type of bonded magnet, the objective is clearly to maximize the volume fraction of magnetic powder that is used, while still maintaining adequate flow characteristics for molding the compound and sufficient particle bonding for good mechanical properties. In fact, it is possible to improve somewhat on the volume fractions by careful attention to the proportions of sizes of particles that are employed, and optimum values for v are around 5% greater than their nominal values, given in Table 2.2.

Because their raw materials are so abundant, by far the most popular type of powder to form into bonded magnets is ferrite. The sintered $BaFe_{12}O_{19}$ material described in Equation (2.18) is milled to a particle size around 1 μm diameter, blended with a binder into a compound, and then molded. In the simplest processes, there is no attempt to align the particles' magnetization vectors, and so the energy products of such magnets are quite low.

Injection molding is frequently used for isotropic ferrites, and while this

Table 2.2. *Volume fractions of magnetic powder in bonded magnet processes.*

	Compression molding	Injection molding	Extrusion
Nominal	80%	60%	55%
Optimum	85%	65%	60%

achieves a typical value for $(BH)_{max}$ of only around 4 kJ/m^3, the absence of any preferred orientation allows often complex pole arrays to be magnetized on the molded part. The simplicity of this process has popularized isotropic injection-molded magnets of various types, though because these compounds have high volume fractions of abrasive magnetic powder, hardened wear-resistant alloys must be used in high-volume production equipment; it is common to modify the screw and barrel of the injection machine in this way, and to use tungsten carbide gate inserts in the mold.

The magnetic properties may be optimized, though, by aligning the particles in the mold, and anisotropic injection-molded ferrites can then achieve a typical $(BH)_{max}$ of 14 kJ/m^3. This is done by applying a field in the mold cavity while the compound is still at its melt temperature and very fluid, though embedding an electromagnetic coil in the mold makes this a much more complex and costly item. Figure 2.22 shows a typical injection mold to produce a radially oriented magnet ring. The sprues and runners that are molded as part of the ring are removed and recycled into the compound to reduce scrap material. A further benefit of this process is that other parts can be molded into the ferrite, by inserting them into the mold before the magnetic material is introduced. An example of this is given in Figure 2.23, which shows a steel shaft that has been insert molded into a ferrite magnet rotor for a small electric motor.

Sintered $BaFe_{12}O_{19}$ particles develop as platelets with their planes perpendicular to the preferred axis in their hexagonal crystal lattice structure, so some degree of magnetic anisotropy can be created in the pressing direction by mechanical pressure alone (this is comparable to the hot plastic deformation process that was described with reference to rapidly quenched $Nd_2Fe_{14}B$ magnets). Mechanical orientation is most effective in parts that are under about 3 mm thick, and so it is usually

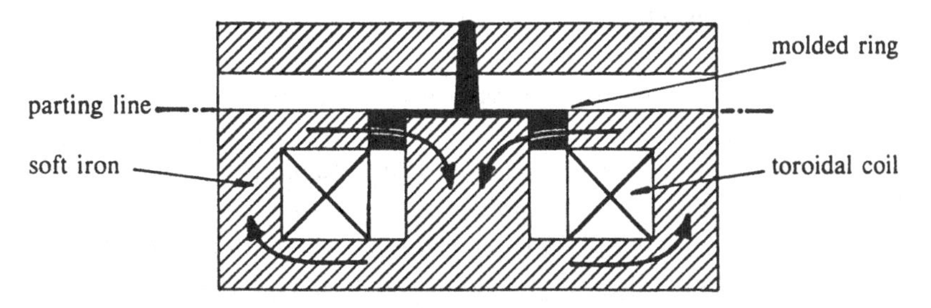

Figure 2.22. Layout of an injection mold for a radially oriented magnet ring.

Figure 2.23. Injection-molded ferrite magnet rotor.

implemented with the extrusion process. Extruded rubber-bonded ferrite magnets can achieve $(BH)_{max}$ through their thickness of around 5 kJ/m^3, but with the added complexity of applying an orienting magnetic field in the same direction, this can be improved up to about 11 kJ/m^3. Extrusion is also used as a convenient blending process for nylon-bonded magnets, after which the material is pelletized and injection-molded into isotropic or anisotropic magnets.

After the discovery of samarium–cobalt magnets, it was immediately apparent that, even with the dilution of these materials in a binder, such bonded magnets would offer significant improvements in energy product compared with conventional fully dense ceramic ferrite and alnico types. While the production process must have regard for the high reactivity of rare earth powders, it is also very desirable to be able to mold directly to the final magnet dimensions. Injection-molding offers the highest volume capability while compression molding achieves the highest energy products, both procedures usually being performed with an orienting field in the cavity to produce an anisotropic magnet.

Selection of the proportions of particle sizes produces a dense packing, with fine particles filling the voids between coarse particles, and this optimizes the magnetic properties via a high remanence. For extrusion and injection molding, typical gate sizes require use of an average particle size around 5–10 μm, which corresponds very well with $SmCo_5$ powder. Whereas $SmCo_5$ is inherently a fine particle *nucleation*-type material, Sm_2Co_{17} is a *pinning*-type magnet with a fine cell structure. Its magnetic properties decline once the average particle size is reduced much below 40 μm (Satoh *et al.*, 1985), such that whereas an injection-molded $SmCo_5$ magnet achieves a $(BH)_{max}$ of around 60 kJ/m^3, one of Sm_2Co_{17} is only slightly better at about 72 kJ/m^3. Sm_2Co_{17} is better suited to compression molding, where its optimum particle size can be used to achieve the best magnetic properties, with a $(BH)_{max}$ as high as 135 kJ/m^3 (Shimoda, 1987). A number of compositional variations are used to give a range of properties for compression-molded Sm_2Co_{17}, the most common being the partial substitution of Fe for some Co, which increases the saturation magnetization while lowering intrinsic coercivity.

Sintered $Nd_2Fe_{14}B$ is a *nucleation*-type material, but when exposed to air, it is highly susceptible to oxidation and its intrinsic coercivity degrades very significantly. For example, it cannot be milled into the 5–10 μm particles required for injection molding, at which size the grain boundaries present most of their surface areas to air, and are much less effective in preventing domain wall motion. Since $Nd_2Fe_{14}B$ has more economical raw materials than Sm_2Co_{17}, bonded magnets made from this generally have to demonstrate advantages in high-volume production to warrant their reduced properties compared with fully dense $Nd_2Fe_{14}B$. This makes injection molding of particular interest, though obviously not using a compound of the sintered powder.

Rapidly quenched $Nd_2Fe_{14}B$, on the other hand, has a very fine magnetocrystalline structure with a grain size around 0.030 μm. It is therefore possible to mill the particles as small as 1 μm with no degradation of its intrinsic coercivity. Even at this size, most of the grain boundaries are not on the particles' surfaces, so the rapidly quenched powder exhibits very little oxidation. The drawback to this material is that it is inherently *isotropic*, but there is no need to provide an orienting field in the mold. Molding with rapidly quenched $Nd_2Fe_{14}B$ powder is therefore about as straightforward as with isotropic ferrite, and almost as popular. HDDR $Nd_2Fe_{14}B$ powder has a grain size around 0.3 μm, and it is also found to be relatively stable due to the limited oxygen pick-up in this process. Isotropic bonded magnets made from either type of powder have similar properties. Typical values for $(BH)_{max}$ that are achieved by compression-molding, injection molding and extrusion are 80, 45 and 36 kJ/m^3 respectively.

Rapidly quenched $Nd_2Fe_{14}B$ powder is formed into a fully dense *anisotropic* magnet using the die-upset process already described. This magnet can be re-ground into a stable powder, and used to form anisotropic bonded magnets by applying an orienting field in the mold.

This whole procedure is complex, but may be warranted when complex final magnet shapes can be molded directly. A simpler procedure is provided by HDDR powder which, with appropriate partial substitutions for Fe, is naturally anisotropic and only requires orienting with a field in the mold. Compression and injection molding achieve energy products of around 190 and 100 kJ/m^3 respectively.

References

Bleaney, B. I. and Bleaney, B. (1965). *Electricity and Magnetism*, 2nd edn. Oxford: The Clarendon Press.

Croat, J. J. (1982). Magnetic hardening of Pr–Fe and Nd–Fe alloys by melt-spinning. *Journal of Applied Physics*, **53**, 3161–8.

Harris, I. R. and McGuinness, P. J. (1990). Hydrogen: its use in the processing of NdFeB-type magnets and in the characterisation of NdFeB-type alloys and magnets. *11th International Workshop on Rare-Earth Magnets and their Applications*, pp. 29–48. Pittsburgh: Carnegie Mellon University.

Heck, C. (1974). *Magnetic Materials and their Applications*, London: Butterworth.

Hopstock, D. M. (1987). Current status of thermoplastic and elastomer bonded neodymium–iron–boron magnets. *9th International Workshop on Rare-Earth Magnets and their Applications*, pp. 667–74. Bad Honnef: Deutsche Physikalische Gesellschaft.

Jones, F. G. (1987). Metallurgical processes for making rare-earth transition metal magnet precursor alloys: an overview. *9th International Workshop on Rare-Earth Magnets and their Applications*, pp. 737–52. Bad Honnef: Deutsche Physikalische Gesellschaft.

Lee, R. W., Brewer, E. G. and Schaffel, N. A. (1985). Processing of neodymium–iron–boron melt-spun ribbons to fully dense magnets. *IEEE Transactions on Magnetics*, **21**, 1958–63.

Livingston, J. D. (1986). Microstructure and properties of rare earth magnets. *Soft and Hard Magnetic Materials with Applications*, pp. 71–9. Metals Park: American Society of Metals.

Satoh, K., Oka, K., Ishii, J. and Satoh, T. (1985). Thermoplastic resin-bonded Sm–Co magnet. *IEEE Transactions on Magnetics*, **21**, 1979–81.

Shimoda, T. (1987). Current situation of bonded rare earth magnets in Japan. *9th International Workshop on Rare-Earth Magnets and their Applications*, pp. 651–65. Bad Honnef: Deutsche Physikalische Gesellschaft.

Strnat, K. J. (1970). The recent development of permanent magnet materials containing rare earth metals. *IEEE Transactions on Magnetics*, **6**, 182–90.

Strnat, K. J. (1987). Permanent magnets based on 4f–3d compounds. *IEEE Transactions on Magnetics*, **23**, 2094–9.

Takeshita, T. and Nakayama, R. (1990). Magnetic properties and microstructures of the Nd–Fe–B magnet powders produced by the hydrogen treatment – (III). *11th International Workshop on Rare-Earth Magnets and their Applications*, pp. 49–71. Pittsburgh: Carnegie Mellon University.

Takeshita, T. and Nakayama, R. (1992). Magnetic properties and microstructures of the Nd–Fe–B magnet powders produced by HDDR process – (IV). *12th International Workshop on Rare-Earth Magnets and their Applications*, pp. 670–81. Perth: University of Western Australia.

3
Thermal stability

3.1 Introduction

When a designer specifies the use of a permanent magnet, he certainly hopes that its magnetization will indeed remain *permanent*, or at least a close approximation to this. Specifically, the designer requires the magnet's *demagnetization curve*, the second quadrant of the *B versus H* characteristic, to remain unchanged under normal operating conditions. Unfortunately, this is never the case, so it is important to understand the nature of the changes that may occur, so that any degradation of the magnetic properties reflected in the demagnetization curve may be accounted for in the design. Changes in a magnet after it has been manufactured and fully magnetized may be caused by any combination of external influences, such as temperature, pressure and applied field. These changes fall into three categories.

The first category comprises those effects that result in a *permanent* change in the demagnetization curve, which persist even if the magnet is fully remagnetized. One should either avoid selecting a particular magnet type for an environment in which it will be exposed to conditions known to cause a permanent change, or provide protection for the magnet from this environment. Consider the case of alnico magnets, which, as described in Chapter 2, undergo a critical segregation of the α_1 and α_2 phases during their heat treatment between 550 and 650 °C. If these magnets are subsequently exposed to operating temperatures above about 500 °C, then the phase composition will be altered; the original demagnetization curve cannot be regained even by full remagnetization, but it can be recovered by re-tempering.

Sintered ferrite and some rare earth magnets suffer no such permanent changes until they reach about 1000 °C, due to their high sintering temperatures, but these materials all have practical limits to their operating temperatures, which are well below this level. All types of rare earth magnet are susceptible to oxidation, which causes a permanent metallurgical change in their structure. The rate and severity of the change will depend on the exact composition and process method, but it is most serious in sintered $Nd_2Fe_{14}B$ magnets, which must be encapsulated for corrosion protection. In Chapter 2, it was described how mechanical pressure is deliberately used to re-order the crystal structure in some rapidly quenched $Nd_2Fe_{14}B$ and ferrite magnets; accidental exposure of a magnet to pressure may of course create an undesirable structural change. A last, notable example of a cause of a *permanent* change in the demagnetization curve is that of the polymer matrix binder employed in all types of bonded magnets, which has a relatively low "heat deflection temperature" above which it becomes pliable.

While one generally tries to avoid selecting a magnet for operating conditions that will expose it to significant permanent changes in its demagnetization curve, it may be common for a magnet to encounter and endure changes from the other two categories. "*Irreversible*" changes are those that persist even after the cause has been removed, but the original demagnetization curve can be restored by fully remagnetizing the material. Depending upon the shape of the demagnetization curve, an externally applied field may cause the magnet to operate on an "internal" curve with a lower $(BH)_{max}$, and dynamic operation of this type is discussed at length in later chapters.

The other principal cause of *irreversible* change in the demagnetization curve is thermal fluctuations. Higher thermal energy can activate reversals of some of the domain magnetizations or movement of the domain boundaries; such effects are sometimes known as *magnetic viscosity*. When the temperature returns to its original value, the "irreversible" loss in magnet energy can only be recovered by full remagnetization. Thermal fluctuations may also cause a less drastic *reversible* change in the demagnetization curve, through agitation of the magnetic moments in the domains. This effect simply reduces the saturation magnetization of the magnet temporarily, until the original temperature is restored.

Further irreversible changes due to thermal effects can be almost completely eliminated by cycling the magnet through a slightly greater temperature range than it will experience in service. This, however, does not alleviate the need to understand degradation in magnet performance that is to be anticipated under its normal operating conditions, nor the nature of reversible changes that will still occur as temperature varies.

3.2 The Curie temperature

The theories developed in earlier chapters do not account for thermal agitation of the magnetic moments, which will disrupt their alignment and reduce the net magnetization $\boldsymbol{M}$. In fact, spontaneous magnetization will not occur at all above a certain temperature, known as the *Curie temperature*, T_c. Clearly, it would be desirable for T_c to be a few times greater than the maximum temperature a material is likely to experience, and yet some of the principal elements used in permanent magnets have rather low values of T_c: 1120 °C for Co, 770 °C for Fe, 358 °C for Ni. Fe is the most popular, economical element, but the ceramic ferrite $BaFe_{12}O_{19}$ has a T_c value of only 450 °C. Complete thermal demagnetization can therefore be achieved by raising ceramic ferrite above its T_c, since permanent metallurgical changes do not occur in this material under about 1000 °C. $Nd_2Fe_{14}B$ magnets may have T_c as low as 300 °C, but partial substitutions such as Co for some of the Fe are used to raise this value somewhat.

Spontaneous magnetization can alone cause a field within a material given by $\boldsymbol{B}=\mu_0\boldsymbol{M}$ (Equation (1.24)). The *magnetization* $\boldsymbol{M}$ in an elemental volume ΔV of this material was defined in Equation (1.7) as

$$\boldsymbol{M}=\lim_{\Delta V\to 0}\frac{\sum \boldsymbol{\mu}_{\mathrm{m}}}{\Delta V} \tag{1.7}$$

This is strictly the magnetic dipole moment ($\boldsymbol{\mu}_{\mathrm{m}}$) per unit volume, which, if the material has m atoms per unit volume, may alternatively be written as

$$\boldsymbol{M}=m\boldsymbol{\mu}_{\mathrm{m}} \tag{3.1}$$

Thermal agitation disrupts the alignment of the moments, an effect quantified using classical statistical mechanics (Jain, 1967). This results in modification of the simple Equation (3.1) relationship to

$$\boldsymbol{M}=m\boldsymbol{\mu}_{\mathrm{m}}\left(\coth\frac{\mu_{\mathrm{m}}B}{kT}-\frac{kT}{\mu_{\mathrm{m}}B}\right) \tag{3.2}$$

where T is the temperature and k is Boltzmann's constant. For high fields or low temperatures, $(\mu_{\mathrm{m}}B/kT)\to\infty$ and $\coth(\mu_{\mathrm{m}}B/kT)\to 1$, so Equation (3.2) reverts to the form of Equation (3.1), representing alignment of all the moments and saturation of the material:

$$\boldsymbol{M}_{\mathrm{sat}}=m\boldsymbol{\mu}_{\mathrm{m}} \tag{3.3}$$

The general relationship for spontaneous magnetization may therefore be written by combining Equations (3.2) and (3.3) and $\boldsymbol{B}=\mu_0\boldsymbol{M}$:

$$\boldsymbol{M}=\boldsymbol{M}_{\text{sat}}\left(\coth\frac{\mu_0\mu_{\text{m}}M}{kT}-\frac{kT}{\mu_0\mu_{\text{m}}M}\right) \tag{3.4}$$

To help in understanding the significance of this relationship, let

$$x=\frac{\mu_0\mu_{\text{m}}M}{kT} \tag{3.5}$$

so that

$$\frac{\boldsymbol{M}}{\boldsymbol{M}_{\text{sat}}}=\coth x-\frac{1}{x} \tag{3.6}$$

$\boldsymbol{M}$ and $\boldsymbol{M}_{\text{sat}}$ are also related to the variable x by combining Equations (3.3) and (3.5) as

$$\frac{\boldsymbol{M}}{\boldsymbol{M}_{\text{sat}}}=\left(\frac{kT}{m\mu_0\mu_{\text{m}}^2}\right)x \tag{3.7}$$

x can now be eliminated between Equations (3.6) and (3.7) to relate $\boldsymbol{M}$ directly to temperature T, but in anticipation of the result, the *Curie temperature*, T_{c}, is defined as

$$T_{\text{c}}=\frac{m\mu_0\mu_{\text{m}}^2}{3k} \tag{3.8}$$

Notice that T_{c} is indeed a fundamental characteristic of the material, being defined only in terms of other fundamental properties and constants: m atoms per unit volume, the constant μ_0, the magnetic dipole moment $\boldsymbol{\mu}_{\text{m}}$, and Boltzmann's constant k. Equation (3.7) then becomes

$$\frac{\boldsymbol{M}}{\boldsymbol{M}_{\text{sat}}}=\frac{T}{3T_{\text{c}}}x \tag{3.9}$$

Spontaneous magnetization can only occur if there is a solution for the simultaneous Equations (3.6) and (3.9), which are each plotted against x in Figure 3.1. The latter is a straight line whose slope, $T/3T_{\text{c}}$, increases with temperature. The curve $(\coth x-1/x)$ approximates to $x/3$ as $x\rightarrow 0$, which proves that $T=T_{\text{c}}$ is the critical temperature for solution of these equations. For $T>T_{\text{c}}$, there is no solution and spontaneous magnetization cannot exist in the material. For $T<T_{\text{c}}$, solutions in x do exist between

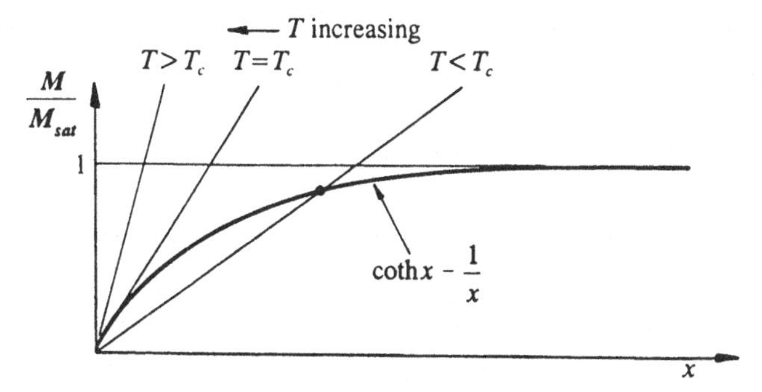

Figure 3.1. Solution for spontaneous magnetization at various temperatures.

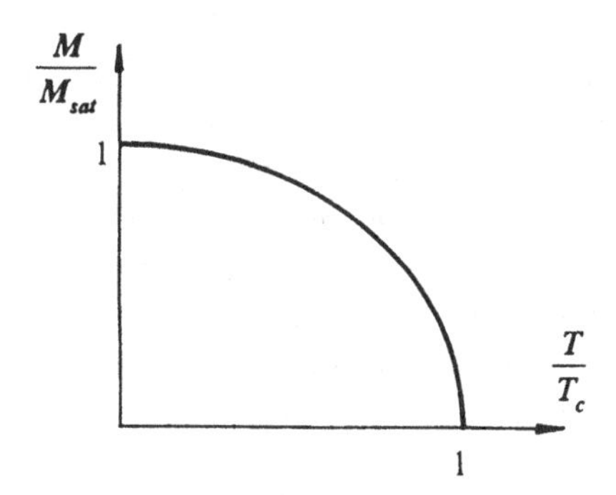

Figure 3.2. Relationship between magnetization and temperature.

Equations (3.6) and (3.9), as illustrated in Figure 3.1, and the relationship between M/M_{sat} and T/T_c, which is thus deduced, is plotted in Figure 3.2.

3.3 Demagnetization curves

A permanent magnet exposed to an elevated temperature will therefore suffer a reduction in its *effective* magnetization relation to the saturation level, in the manner of Figure 3.2. Whether this reduction is *reversible* or *irreversible* will depend upon other factors, most notably the magnetizing force within the material. When this reaches the intrinsic coercivity level, $-H_{ci}$, the magnetization changes to the opposite direction - an "irreversible" change, from which the original condition can be restored by remagnetization with a field exceeding $+H_{ci}$. Provided that the magnetizing force is contained within the range $-H_{ci}<H<+H_{ci}$, any consequent changes in the magnetization will be "reversible". It is obviously important to know whether the magnetizing force exceeds the magnitude of H_{ci}, but this parameter itself is a function of temperature.

The intrinsic coercivity for a material with *magnetocrystalline anisotropy* was found in Equation (1.40) to be

$$H_{ci} = \frac{2K_1}{\mu_0 M_{sat}} \tag{1.40}$$

K_1 is a crystallographic constant of the material, written using Equation (3.1) as $8\mu_0 m\mu_m^2$. As temperature increases and the *effective* magnetization is reduced, Equation (1.40) shows that H_{ci} will increase. The "ideal" intrinsic magnetization characteristic for magnetocrystalline anisotropy in Figure 1.10, and the *B versus H* loop derived from it in Figure 1.11, both exhibit abrupt changes as prescribed by the theory at $\pm H_{ci}$. Real magnets, on the other hand, have gradual transitions and there is said to be a *knee* in the *demagnetization curve*, the second quadrant of the *B versus H* loop. A family of demagnetization curves can therefore be plotted, each for the material at a different operating temperature.

Ceramic ferrites base their permanent magnetism on magnetocrystalline anisotropy, and a typical set of demagnetization curves for the Ceramic 8 grade is shown in Figure 3.3. Ceramic 8 is a common anisotropic sintered magnet with a maximum energy product around 28 kJ/m^3. At a given temperature, if $|H_{ci}| > M_{sat}$, then the demagnetization curve (in the

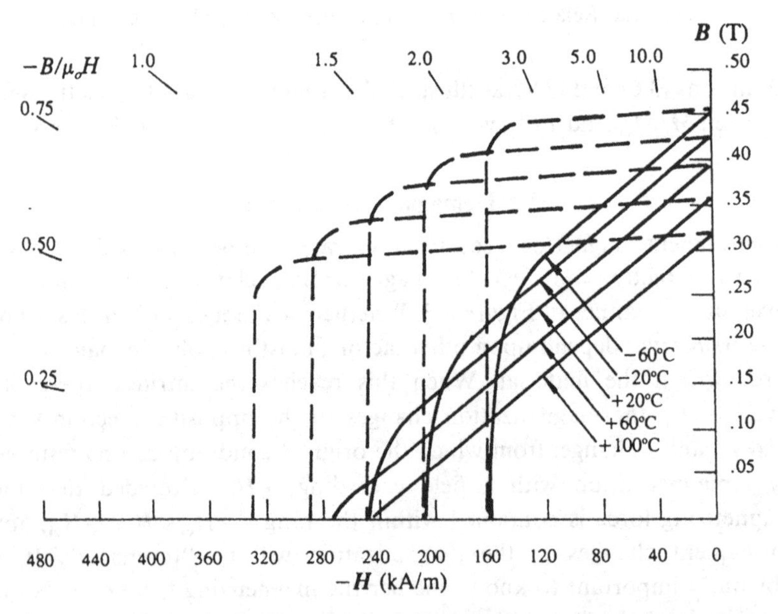

Figure 3.3. Demagnetization curves of Ceramic 8 at various temperatures.

second quadrant) is entirely linear and the knee at $-H_{ci}$ is moved out into the third quadrant of the B *versus* H loop. This condition appears to occur for Ceramic 8 when it is elevated above room temperature. When it is cooled, though, the magnetization is restored with a consequent rise in the remanence B_r, while H_{ci} is reduced. A transition occurs for this material at around room temperature, below which $|H_{ci}|<M_{sat}$ and the knee at $-H_{ci}$ moves into the second quadrant demagnetization curve. This sequence is peculiar to materials that are dominated by magneto-crystalline anisotropy, such as the ceramic ferrites.

Figure 3.3 shows that operation of a magnet at a point beyond the knee of a demagnetization curve ($H<-H_{ci}$) represents a reversal of magnetization within the material, an *irreversible* change. Consider a magnet whose operating condition varies *only* in respect of its temperature, change in which causes the demagnetization curve to alter in the manner of Figure 3.3. To simplify matters, assume that the magnet experiences a *uniform* internal field, which may therefore be expressed by Equations (1.52) and (1.53):

$$\boldsymbol{H}_i=\boldsymbol{H}_o-N\boldsymbol{M} \tag{1.52}$$

$$\boldsymbol{B}_i=\mu_0\boldsymbol{M}(1-N) \tag{1.53}$$

The operating point of a magnet on its demagnetization curve must also be defined by a *load line*, whose slope is derived from these equations as

$$\frac{\boldsymbol{B}_i}{\mu_0\boldsymbol{H}_i}=\frac{\boldsymbol{M}(1-N)}{\boldsymbol{H}_o-N\boldsymbol{M}} \tag{3.10}$$

The *demagnetizing factor* N depends only on the magnet's dimensions and is a constant, as is the external field $\boldsymbol{H}_o$. The slope of the load line will therefore not change provided that $\boldsymbol{M}$ is constant, and there will be a unique operating point for the magnet defined by its intercept with the demagnetization curve.

Some curves from Figure 3.3 are redrawn in Figure 3.4, together with an example load line representing $\boldsymbol{B}_i/\mu_0\boldsymbol{H}_i\approx-1$. At $+20\,°\mathrm{C}$, the operating point $\boldsymbol{a}$ is well above the knee. When the temperature falls to $-20\,°\mathrm{C}$, the magnet shifts to operating point $\boldsymbol{b}$, still above the knee of that curve. Actually $\boldsymbol{M}$ is slightly higher now, but the small change in the load line slope will be ignored for clarity in this example. Continued operation between $+20\,°\mathrm{C}$ and $-20\,°\mathrm{C}$ is *reversible*, as illustrated by the inset in Figure 3.4, though the flux density in the magnet increases over this range as temperature falls.

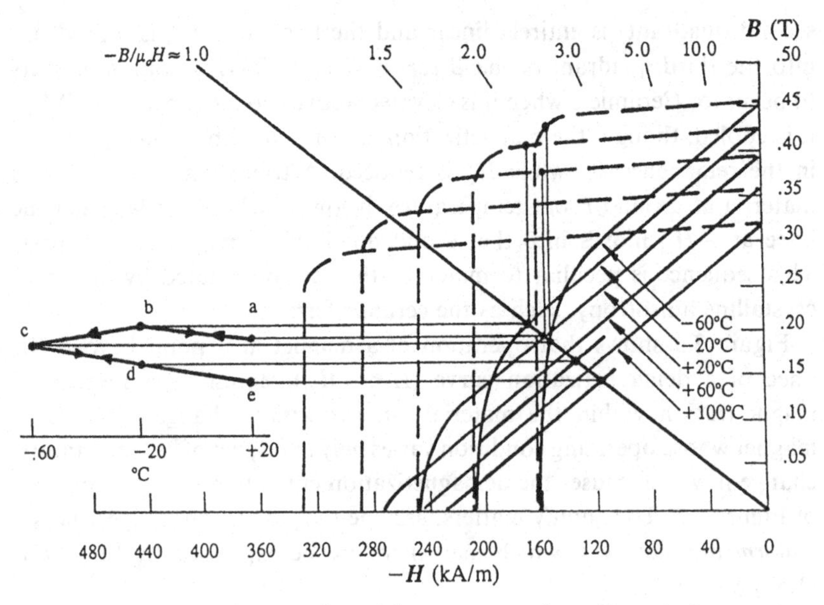

Figure 3.4. Change in operation with temperature for a Ceramic 8 magnet.

If the temperature falls further to $-60\,°C$, the magnet moves to operating point ***c***, which is just *below* the knee of the $-60\,°C$ curve. There is a *decrease* in magnetization and flux density, and this change is now *irreversible*. It is clear from Figure 3.4 that even a slight increase in the demagnetizing field $-\boldsymbol{H}_i$ will result in a catastrophic loss of $\boldsymbol{M}$. Since complete alignment of $\boldsymbol{M}$ can only be restored with full remagnetization of the material, a return to operation at higher temperatures such as $-20\,°C$ and $+20\,°C$ will not be accompanied by a return to the corresponding "major" demagnetization curves. Operating points ***d*** at $-20\,°C$ and ***e*** at $+20\,°C$ will lie at the intersections of the load line with "minor" demagnetization curves within the "major" *B versus H* loop, and represent reduced operating flux densities. Further cycling in the temperature range $-60\,°C$ to $+20\,°C$ will now be *reversible*, and the magnet may be considered to be stabilized at a reduced energy.

The changes that occur in a magnet are also dependent upon its *shape*, and its dimensions may be used to control the excursions of the operating point with temperature. A popular parameter is the L/D ratio, which applies to a circular disk of diameter D, which is magnetized along its length L. For a very long, thin rod, magnetized along its major axis, $L/D \to \infty$, the demagnetizing factor $N \approx 0$, and the slope of the load line increases according to Equation (3.10) to $\boldsymbol{B}_i/\mu_0 \boldsymbol{H}_i = \boldsymbol{M}/\boldsymbol{H}_o$. In the case of

Ceramic 8 operating between $-60\,°C$ and $+20\,°C$, Figure 3.4 shows that a sufficient increase in load line slope may be used to convert the *irreversible* change into a *reversible* change between these temperature limits. Equation (1.53) shows that $N \approx 0$ is the limit in which self-demagnetizing field is absent and spontaneous magnetization most easily exists. At the other extreme, there is a strong self-demagnetizing field, which prevents spontaneous magnetization when $N \approx 1$, as in a very thin disk ($L/D \rightarrow 0$) magnetized normal to its plane. The load line slope falls ($\boldsymbol{B}_i/\mu_0\boldsymbol{H}_i \rightarrow 0$), which in the example shown in Figure 3.4 causes Ceramic 8 to exhibit *irreversible* changes for *all* temperatures below $+20\,°C$.

For each permanent magnet material, there is a critical L/D (or equivalent) ratio below which *reversible* changes over a given temperature range will become *irreversible*. A set of curves may be constructed as with the inset of Figure 3.4 to illustrate the temperature dependence of flux density at various operating conditions, each characterized by a unique load line of slope $\boldsymbol{B}_i/\mu_0\boldsymbol{H}_i$. The curves for Ceramic 8 are shown in Figure 3.5, which indicate the onset of irreversible operation at progressively lower temperature threshholds as $\boldsymbol{B}_i/\mu_0\boldsymbol{H}_i$ gets larger.

The intrinsic coercivity of alnico magnets is dominated by *shape anisotropy*, which was given in Equation (1.60) as

$$H_{ci} = M_{sat}(N_b - N_a) \tag{1.60}$$

As the *effective* magnetization declines with increasing temperature, so too will the component of H_{ci} due to shape anisotropy. Determining the temperature dependence of alnicos is more complicated than with ferrites, though. Magnetocrystalline anisotropy is still present, albeit in a secondary role, and the characteristics of an alnico magnet will depend upon the exact constituents of the alloy; cobalt, for example, being included to improve the anisotropy. Nevertheless, Equation (1.60) suggests that a steady M_{sat} will stabilize H_{ci}, and alnico magnets in general do exhibit much smaller changes in their magnetic performance with temperature than ferrites. The demagnetization curves of Alnico 5 grade are shown in Figure 3.6, from which the temperature dependence of flux density at various operating conditions may be derived as before. Because alnicos inherently have $|H_{ci}| < M_{sat}$, it is usual for higher load line slopes to be employed than with ferrites. The results in Figure 3.7 are similar in form to those in Figure 3.5, except that the percentage change in flux with temperature is much smaller in the Alnico 5 material.

Once a magnet has been cycled through the normal temperature range it will experience in service, there will be no further *irreversible* changes

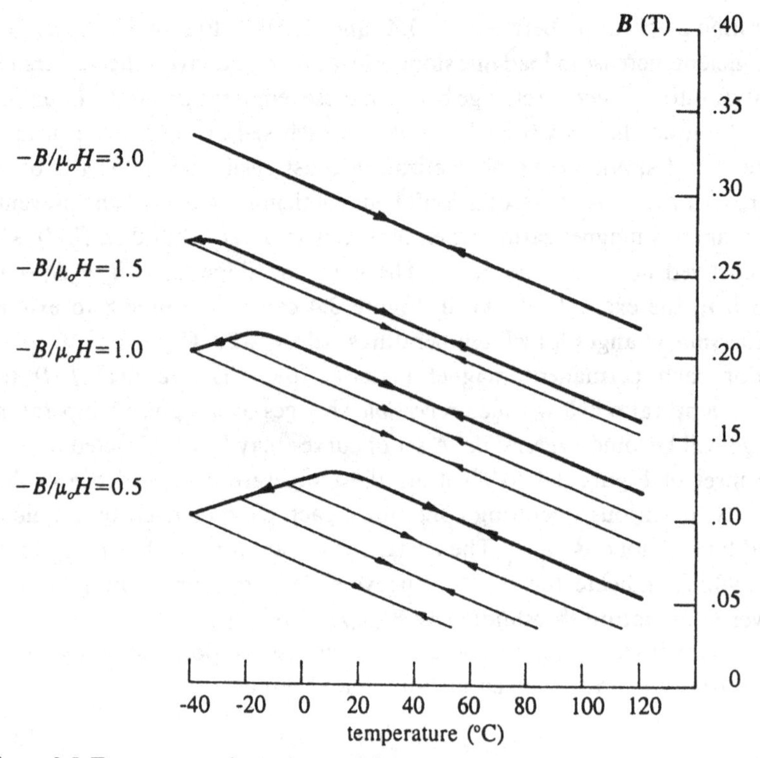

Figure 3.5. Temperature dependence of Ceramic 8 with load lines of various slopes.

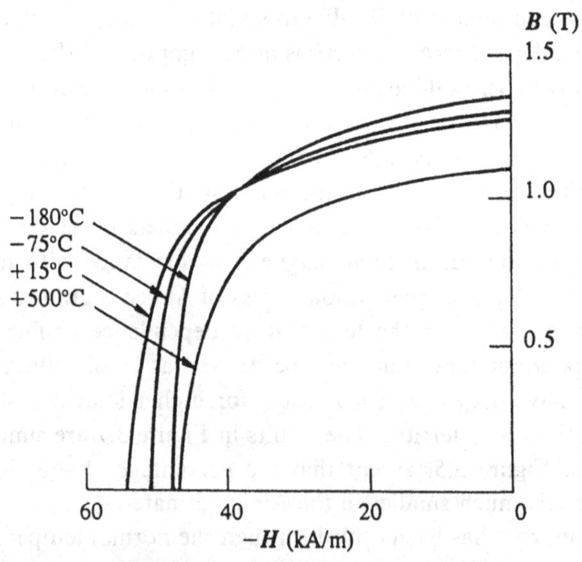

Figure 3.6. Demagnetization curves of Alnico 5 at various temperatures.

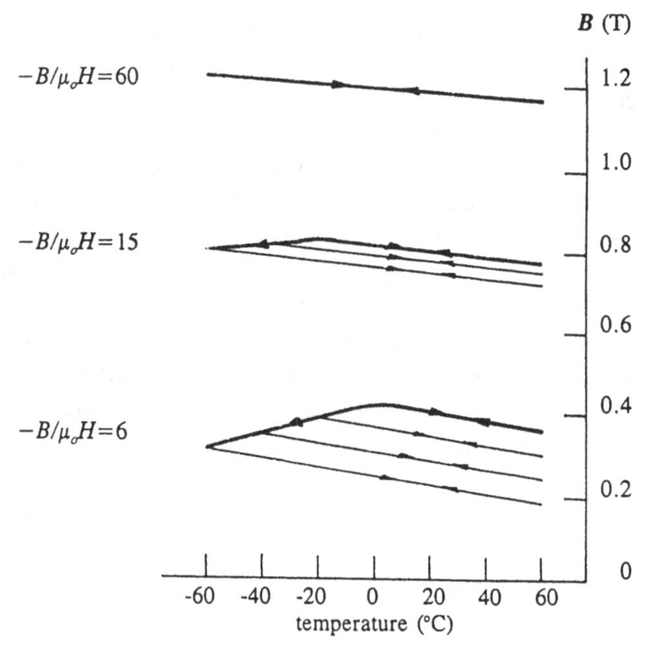

Figure 3.7. Temperature dependence of Alnico 5 with load lines of various slopes.

due to thermal effects, and it will operate along a *reversible* characteristic. The examples given in Figures 3.5 and 3.7 indicate that, for a particular material, it is approximately valid to assign a specific value to the *reversible* percentage change in magnetic flux per unit change in temperature, irrespective of the original temperature excursion or the magnet's shape, i.e. the load line slope. This commonly used parameter of permanent magnets is called the *reversible temperature coefficient* (α), having values of about $-0.2\%/°C$ for Ceramic 8 and $-0.02\%/°C$ for Alnico 5.

There will obviously be an upper temperature limit at which this change in magnet flux is no longer *reversible*. Although sintered ferrite magnets suffer no *permanent* changes until they reach about 1000 °C, spontaneous magnetization is destroyed at their T_c of 450 °C, a change that we have identified as *irreversible*. In a manner akin to passing beyond the knee of the demagnetization curve, Figure 3.2 showed that there is some temperature ($< T_c$) above which a substantial loss of magnetization will occur, and full recovery may only now be achieved by returning to a lower temperature *and* by full remagnetization. The situation with alnicos is more serious, even though they have high Curie temperatures in the range 700–850 °C. There will be a *permanent* change in the phase composition if these magnets are exposed to temperatures above about 500 °C, and full recovery to the original condition can only be accomplished by reprocessing the magnet.

A measurement of magnet flux over a temperature range will indicate the nature of its change, as illustrated in Figure 3.8. For a given load condition, the flux change is *reversible* up to some temperature T_1, although operating point ***b*** lies on a demagnetization curve (shown inset) that represents a reduced magnet energy. If the temperature increases further to T_2, the magnet shifts to operating point ***c***, but now an *irreversible* loss of magnet energy and flux has occurred – a return to the original temperature results in operation on a "minor" demagnetization curve at point ***d***. Full remagnetization of the magnet will restore operation to point ***a***, and recover all the lost flux. However, if a *permanent* change has occurred at T_2, such as an alteration to the phase composition, then operation on the original "major" demagnetization curve cannot be restored by remagnetization, and this will only raise the flux from point ***d*** to point ***e***, which is also on a curve representing reduced magnet energy. Upon repeated cycling to a temperature such as T_2, small additional *irreversible* or *permanent* changes may occur, so as a practical matter it is recommended that a magnet be cycled several times through its anticipated operating reversible range.

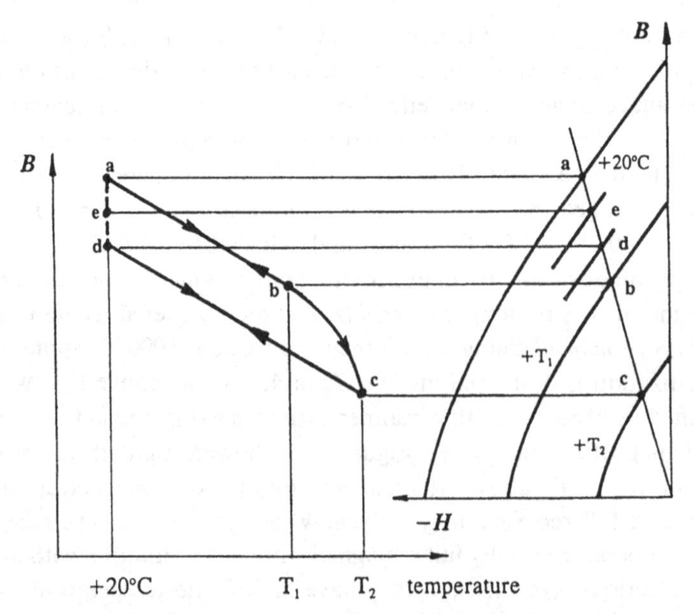

Figure 3.8. Change in magnet flux density with temperature.

3.4 Composition variations

The thermal stability of ceramic ferrites and alnicos has been explained with reference to their anisotropy. As described in Chapter 2, rare earth magnets have high uniaxial magnetocrystalline anisotropy, but this mechanism is more complex than with ferrites because it may also involve nucleation or pinning. Samarium–cobalt magnets have relatively high Curie temperatures, in the ranges 500–750 °C for $SmCo_5$ and 780–850 °C for Sm_2Co_{17}, so the decline in magnetization predicted by Figure 3.2 is slightly less for Sm_2Co_{17} compounds than for $SmCo_5$.

The intrinsic coercivities in these and other light rare earth compounds do not, however, *increase* as M_{sat} declines, but rather H_{ci} *falls* with increasing temperature. This is not specifically due to nucleation or pinning complicating the magnetocrystalline anisotropy mechanism; the cause is at the atomic level, where the rare earth magnetic moments themselves are *strongly* temperature dependent. Because it is proportional to μ_m^2, the crystallographic *constant* K_1 is no longer invariant, but will control the change in H_{ci} via Equation (1.40). Demagnetization curves for the two samarium cobalt materials are shown in Figure 3.9, together with their intrinsic M *versus* H characteristics, which show the decline of H_{ci} with temperature. At least up to 200 °C, it is apparent that $|H_{ci}| > M_{sat}$ for

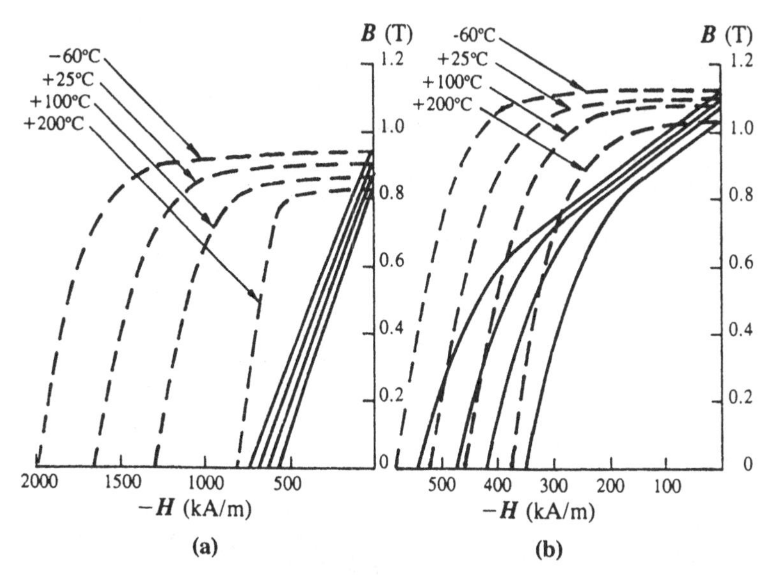

Figure 3.9. Demagnetization curves of sintered magnets at various temperatures, (*a*) $SmCo_5$, (*b*) Sm_2Co_{17}.

$SmCo_5$, and its demagnetization curves are approximately linear throughout this second quadrant. The change in μ_m, K_1 and hence H_{ci} with temperature is more severe in Sm_2Co_{17}, such that $|H_{ci}| \leqslant M_{sat}$ and there is a *knee* in each demagnetization curve. According to these measurements, the knee occurs at higher flux densities as temperature rises, further aggravating the problems of application of this magnet.

To avoid an *irreversible* loss of magnet energy with Sm_2Co_{17} over the temperature range given in Figure 3.9, it appears that it must operate with a load line slope of $\boldsymbol{B}_i/\mu_0\boldsymbol{H}_i > 5$. Figure 3.10 shows the loss in flux when this magnet is cycled up to 200 °C, the *irreversible* change being greater as the load line slope decreases, though the subsequent *reversible* change is almost independent of these conditions. The *reversible temperature coefficient* (α) for Sm_2Co_{17} is about −0.03%/°C, slightly better than that of $SmCo_5$, which is about −0.045%/°C over the same temperature range. Rare earth magnets with more stable properties can be made with partial substitutions of the elements.

There are other heavy rare earth–cobalt compounds using Gd, Tb, Dy, Ho and Er, which exhibit a different temperature dependence of their moments because of a different mode of coupling between the magnetic

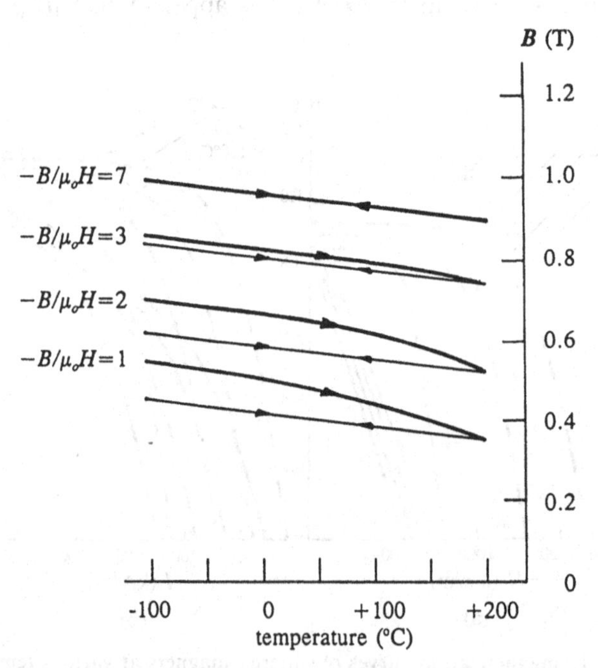

Figure 3.10. Change in flux density with temperature for sintered Sm_2Co_{17}.

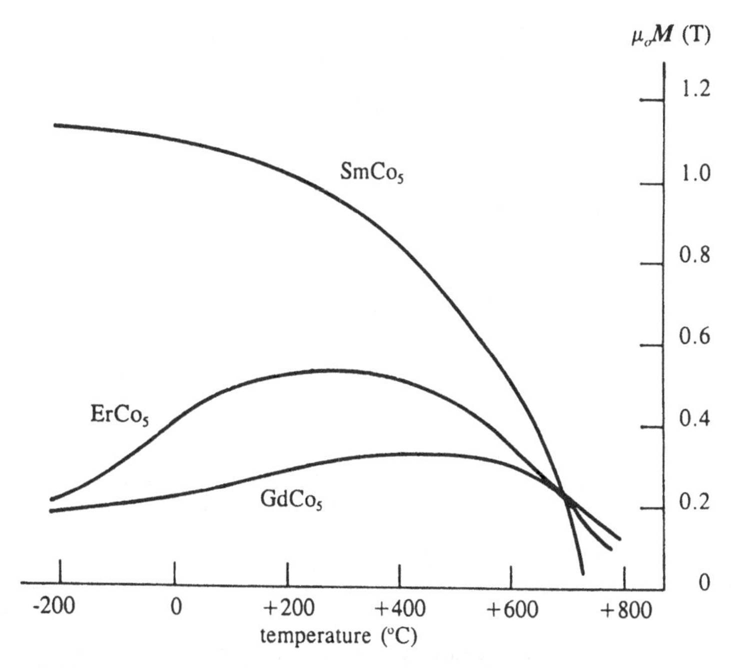

Figure 3.11. Relationship between magnetization and temperature for light and heavy rare earth–cobalt materials.

atoms (Narasimhan, 1981). For example, whereas the temperature dependence of magnetization in $SmCo_5$ follows the form of Figure 3.2, those characteristics for $ErCo_5$ and $GdCo_5$ (Figure 3.11) have *positive* slopes over the temperature range of interest. By combining both light and heavy rare earth elements in the compound, it is possible to achieve a very low reversible temperature coefficient over a limited temperature range. Because the magnetization is much lower with the heavy elements, the improvement in temperature stability is achieved with a corresponding loss in magnet energy product. For example, whereas α for $Sm_{0.8}Gd_{0.2}Co_5$ is improved to $-0.0007\%/^\circ C$, its energy product is only 80 kJ/m^3, half that of a $SmCo_5$ magnet.

Because it starts with a higher energy product, there has been more interest in adapting the Sm_2Co_{17} compound for better temperature stability. With adjustment of the composition and the post-sintering process, a $Sm_{0.6}Er_{0.4}(Co, Fe, Cu, Zr)_{7.22}$ magnet achieves an α of $-0.004\%/^\circ C$ at a $(BH)_{max}$ of 130 kJ/m^3 (Leupold *et al.*, 1984), and a $Sm_{0.6}Gd_{0.4}(Co, Fe, Cu, Zr)_{7.4}$ has an α of $-0.005\%/^\circ C$ with $(BH)_{max}$ of 140 kJ/m^3 (Wang and Chang, 1987). However, some commercial magnets

Table 3.1. *Temperature stability of "Sm_2Co_{17}" magnets.*

$(BH)_{max}$ (kJ/cm^3)	Reversible temperature coefficient (%/°C)
240	−0.030
140	−0.010
130	−0.001

have even lower coefficients than this. The relationship between loss of energy product and decrease in reversible temperature coefficient is highly non-linear, as illustrated in Table 3.1.

While composition variations are used in samarium–cobalt to produce magnets with high temperature stability for some important but specialized applications, they are used as a matter of course in neodymium–iron–boron compounds. The Curie temperature of $Nd_2Fe_{14}B$ is only around 310 °C, and the strong temperature dependence of the magnetic moment of neodymium means that, as temperature increases, there is not only a rapid drop in the magnetization, but an even faster decline in H_{ci} to zero at about 250 °C. Without partial substitutions for Nd and/or Fe, $Nd_2Fe_{14}B$ cannot be used much above 100 °C without suffering a substantial loss of flux, and even over this range its reversible temperature coefficient $\alpha = -0.15\%/°C$, five times greater than that for Sm_2Co_{17}.

We have noted that $Dy_2Fe_{14}B$ has a higher anisotropy than $Nd_2Fe_{14}B$, though with a lower magnetization and hence lower maximum energy product; it also has $\alpha \approx -0.02\%/°C$. A partial substitution of the heavy rare earth dysprosium for Nd significantly increases H_{ci} as shown in Figure 3.12, and reduces α. It has also been noted that cobalt-based compounds have much higher Curie temperatures than those with iron, but lower intrinsic coercivities. A partial substitution of Co for Fe is therefore used to raise T_c in accordance with Figure 3.13. Improvements in both α and T_c are achieved by introducing Dy and Co *together*, in relative amounts chosen also to offset the contrary effects these elements have on H_{ci}. Typically, T_c is raised to around 500 °C and α is reduced to as low as $-0.07\%/°C$. Nevertheless, the antiferromagnetic coupling between Dy and Co reduces the magnetization and $(BH)_{max}$ of the more stable alloy. Demagnetization curves for a commercial sintered (Nd, Dy)–Fe–B magnet in Figure 3.14 (corresponding to those in Figure 3.12) illustrate that high H_{ci} is achieved with a reduced $(BH)_{max}$ of 240 kJ/m^3. The (Nd, Dy)–

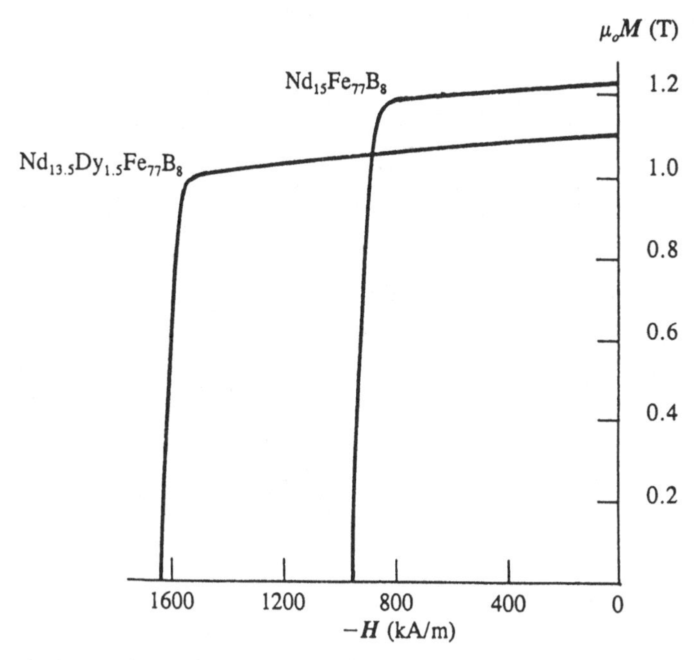

Figure 3.12. Intrinsic demagnetization curves for (Nd, Dy)–Fe–B at room temperature.

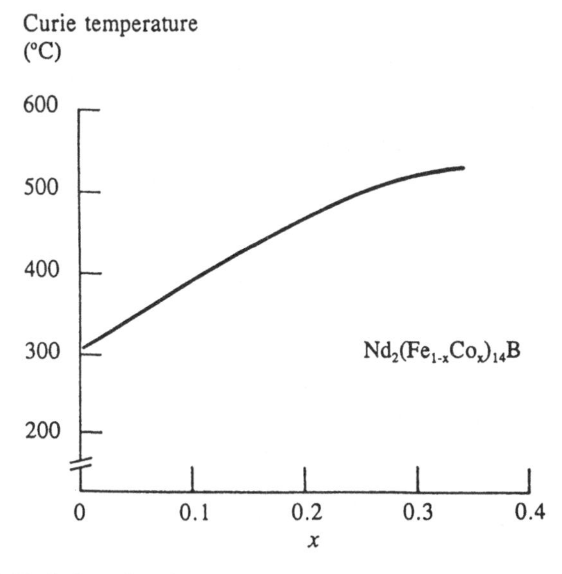

Figure 3.13. Variation of Curie temperature with Co content in $Nd_2(Fe_{1-x}Co_x)_{14}B$.

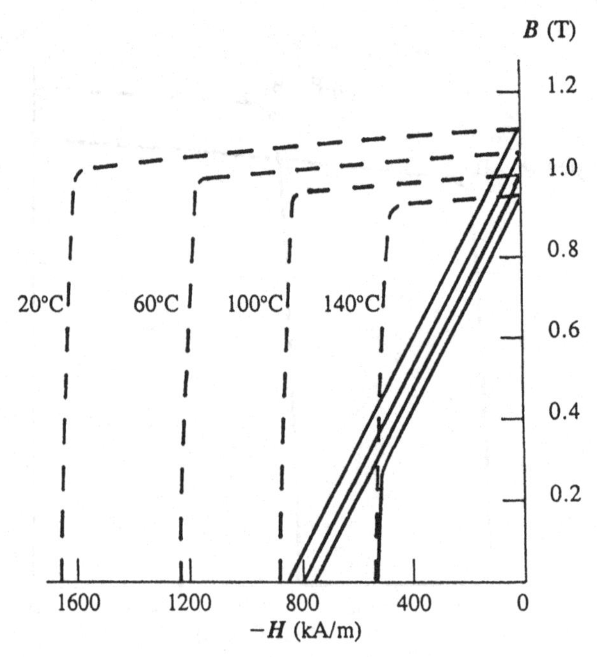

Figure 3.14. Demagnetization curves of (Nd, Dy)–Fe–B at various temperatures.

(Fe, **Co**)–B of Figure 3.15 has improved thermal stability, but with a $(BH)_{max}$ of 280 kJ/m^3, which is still somewhat below the 320 kJ/m^3 potential level for the $Nd_2Fe_{14}B$ compound. Also, the lower H_{ci} due to the Co content may restrict the choice of load line slope to ensure "dynamic" stability over the magnet's operating range.

An equivalent $Nd_2Fe_{14}B$ magnet, which is made using rapidly quenched powder, exhibits similar thermal stability characteristics to the sintered type. Intrinsic demagnetization curves of a fully dense die-upset magnet (Figure 3.16(*b*)) are comparable to those for sintered $Nd_2Fe_{14}B$ (Figure 3.15). However, prior to gaining its strong anisotropy through alignment of the grains' magnetizations in the die-upsetting process, the fully dense hot-pressed magnet is *isotropic* with a lower *net* magnetization (Figure 3.16(*a*)) (Lee, Brewer and Schaffel, 1985). Higher H_{ci} at any given temperature is due to the *shape* of the platelets as described in Section 2.6 rather than to any compositional variations. Nevertheless, higher H_{ci} in isotropic rapid quench powder magnets does lead to improved "dynamic" stability.

The sensitivity of $Nd_2Fe_{14}B$ to temperature has led to the development of a wide range of magnets with various composition variations to provide

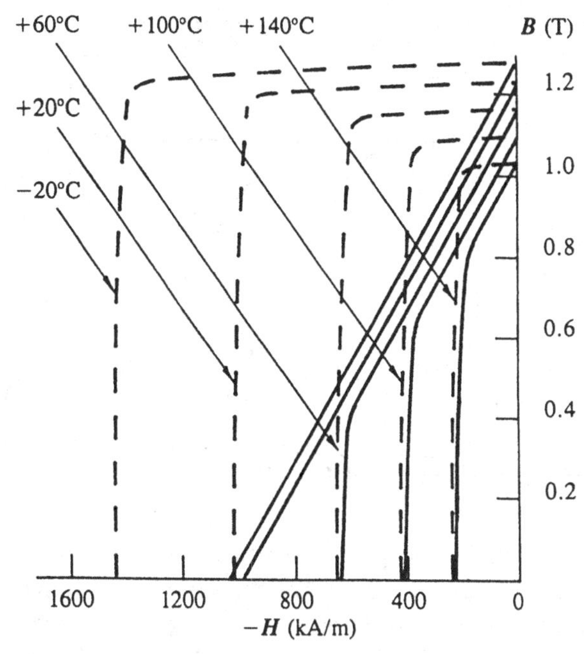

Figure 3.15. Demagnetization curves of (Nd, Dy)–(Fe, Co)–B at various temperatures.

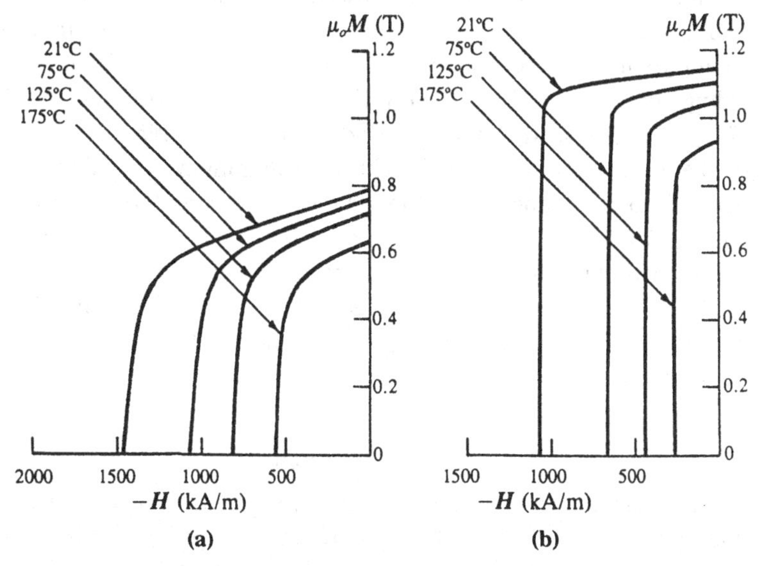

Figure 3.16. Intrinsic demagnetization curves of rapidly quenched $Nd_2Fe_{14}B$, which are (*a*) hot-pressed and then (*b*) die-upset.

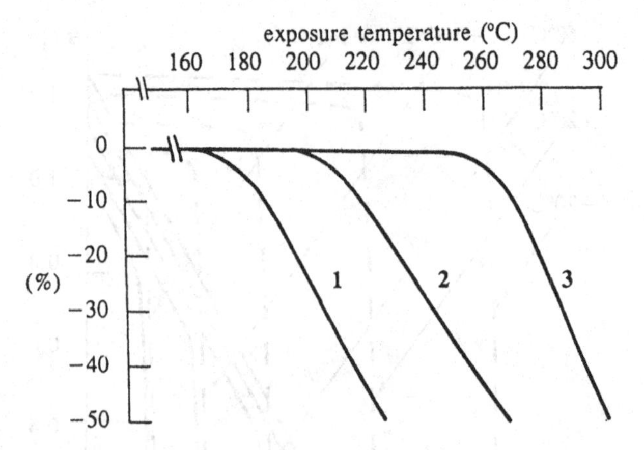

Figure 3.17. Irreversible loss of flux with temperature for 1: (Nd, Dy)–Fe–B, 2: (Nd, Dy)–(Fe, Co)–B, and 3: (Nd, Dy)–(Fe, Co, Nb, Ga)–B.

properties to suit differing applications. Small additions of several different elements are quite common, just a few examples being given here. Gallium (Ga) has been found to improve the alignment capability of the compound, such that H_{ci} is improved without significantly sacrificing magnetization or $(BH)_{max}$ (Tokunaga, Endoh and Harada, 1987). A small amount of niobium (Nb) improves the "squareness" of the demagnetization curve. Figure 3.17 illustrates the temperature to which a magnet may be exposed before it experiences any significant *irreversible* loss in flux, equivalent to moving beyond point ***b*** towards point ***c*** in Figure 3.8. According to Figure 3.17, (Nd, **Dy**)–Fe–B may be exposed to 160 °C, (Nd, **Dy**)–(Fe, **Co**)–B to 200 °C, but (Nd, **Dy**)–(Fe, **Co**, **Nb**, **Ga**)–B may be raised almost to 260 °C. A small addition of aluminum (Al) will increase H_{ci} around room temperature, though not at elevated temperatures. Vanadium (V) has been found to substantially increase the intrinsic coercivity without affecting the saturation magnetization, and a $Nd_{13.5}Dy_{1.5}Fe_{67.5}Co_5V_4B_8Al_{0.5}$ magnet with a room temperature H_{ci} of around 1900 kA/m offers a much improved "dynamic" stability (Tenaud, Vial and Sagawa, 1990).

3.5 Surface oxidation

Although ceramic ferrite magnets have a high *reversible temperature coefficient*, their surfaces are stable and are not susceptible to oxidation. Alnico magnets have a low α, and are also highly resistant to oxidation. Oxidation is, however, a serious problem in many rare earth magnets, dominated by the corrosion that occurs at their surfaces. If no coating is provided for protection, oxygen diffuses into such a magnet, causing a metallurgical change in a surface layer. It has been found that this layer is quite distinctive, with a depth d_0, which is a function of both temperature T and time t (Adler and Marik, 1981):

$$d_0 = \text{fn}(T) \cdot \sqrt{t} \tag{3.11}$$

Typical surface oxidation is shown in Figure 3.18, for a $SmCo_5$ magnet. At elevated temperatures, the product of oxidation in samarium–cobalt magnets is mostly Sm_2O_3.

The consequence of oxidation is that the surface layer possesses a lower H_{ci}, which will allow this region of the magnet to be demagnetized more easily. A *thin* magnet will have a relatively large surface area relative to its volume, and will therefore experience worse oxidation. Unfortunately, the high coercivity of rare earth magnets usually means that they are thin by design. In Figure 1.13, we showed that a magnet's own internal field $\mu_0 \boldsymbol{H}$ is *opposed* to its magnetization $\mu_0 \boldsymbol{M}$, so the degradation in performance due to surface oxidation will be more severe in a magnet

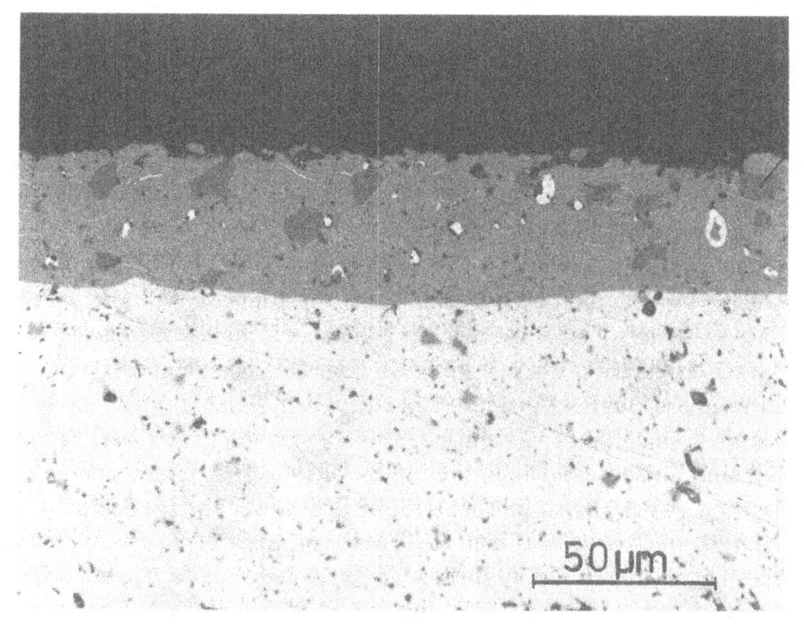

Figure 3.18. Oxidized surface layer on a $SmCo_5$ magnet. (Courtesy of Vacuumschmelze GmbH.)

operating in a long air gap. An example of this is a free-standing magnet, which is said to be operating *open circuit*. The raw powder from which a magnet is made will be highly susceptible to oxidation – d_0 can quickly become comparable to the particle size, so it is important to process magnets from powder without delay. Magnets that are not fully dense, such as many bonded types, will contain porosity, which will allow corrosion to occur at these sites *within* the volume.

Because fn (T) in Equation (3.11) is highly non-linear, there is a critical temperature above which uncoated magnets experience noticeable degradation in magnetic performance, and below which it is safe to operate them. This temperature is around 150 °C for fully dense $SmCo_5$ and around 250 °C for Sm_2Co_{17}. Additional cobalt in the compound reduces the effects of oxidation. Conversely, the lack of cobalt in neodymium–iron–boron makes surface oxidation a much more severe problem in these magnets, and while it is common to provide a protective coating, corrosion resistance is also strongly dependent upon the exact alloy composition. At elevated temperatures, the product of oxidation in Nd–Fe–B is mostly Nd_2O_3 (Jacobson and Kim, 1987). At high humidity, Nd–Fe–B also reacts with hydrogen in the atmosphere, which is absorbed into the surface layer, causing it to disintegrate (Willman and Narasimhan, 1987). These magnets are actually made from Nd-rich powders such as $Nd_{15}Fe_{77}B_8$, such that highly magnetic grain interiors of $Nd_2Fe_{14}B$ are surrounded by Nd-rich grain boundaries. The boundaries provide pinning of the domain walls, and are the source of the magnet's coercivity. Figure 3.19 shows that corrosion progresses from the surface selectively at the grain boundaries, and as this structure breaks down, so too will H_{ci} in the surface layer (Ohashi *et al.*, 1987).

In the previous Section, partial substitution of Co in $Nd_2Fe_{14}B$ was noted to raise T_c, and, even in small concentrations, it will also reduce surface oxidation. This is because the added Co is mainly segregated into the grain boundaries, where it forms an intermetallic compound with the Nd-rich phase. Since it is oxidation of Nd, rather than Co, which modifies the grain boundaries promoting corrosion, formation of this intermetallic compound directly hinders formation of Nd_2O_3. The oxygen content in $Nd_{15}(Fe_{1-x}Co_x)_{77}B_7Al_1$ powder is found to be lower and the degradation of H_{ci} very much reduced (Figure 3.20) as the proportion of Co is increased (Ohashi *et al.*, 1987). The addition of V was noted to raise H_{ci}, but it also reduces corrosion because it, too, inhibits oxidation of the Nd-rich phase in the grain boundaries. $Nd_{13.5}Dy_{1.5}Fe_{67.5}Co_5V_4B_8Al_{0.5}$ appears to be stable even in high humidity up to around 200 °C (Tenaud *et al.*, 1990).

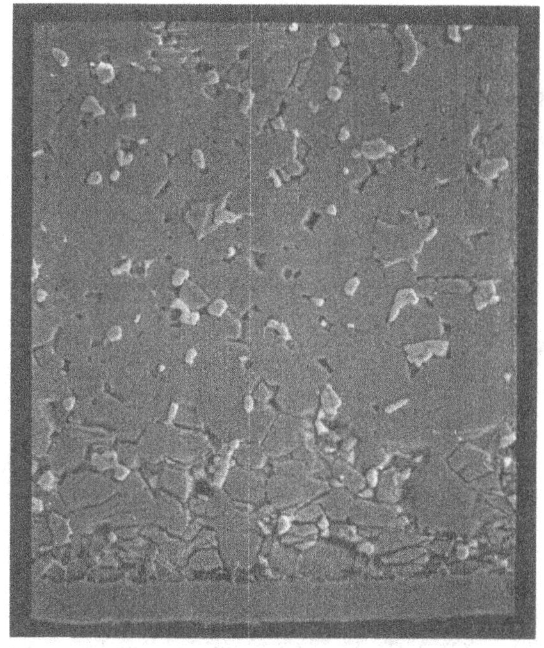

Figure 3.19. Corrosion in a Nd–Fe–B magnet, beneath the Ni-plated surface. (Courtesy of Shin-Etsu Chemical Co., Ltd.)

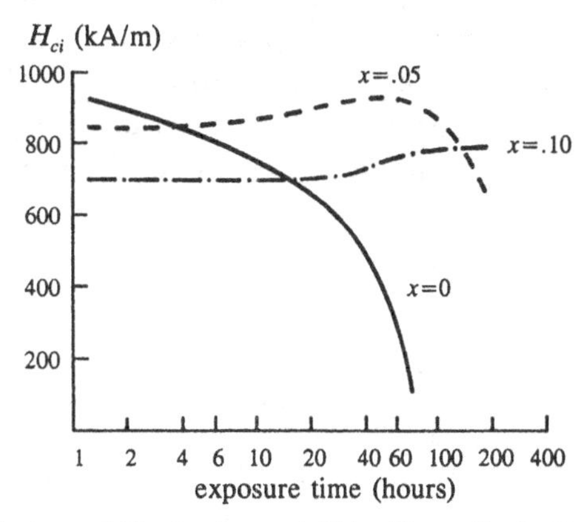

Figure 3.20. Variation of H_{ci} in sintered $Nd_{15}(Fe_{1-x}Co_x)_{77}B_7Al_1$ with various cobalt proportions, x.

The effects of corrosion upon performance of Nd–Fe–B magnets are obviously serious, and degradation of magnetic properties can only be truly contained by both controlling the composition and coating the finished magnet. To be effective, the coating layer should have no porosity,

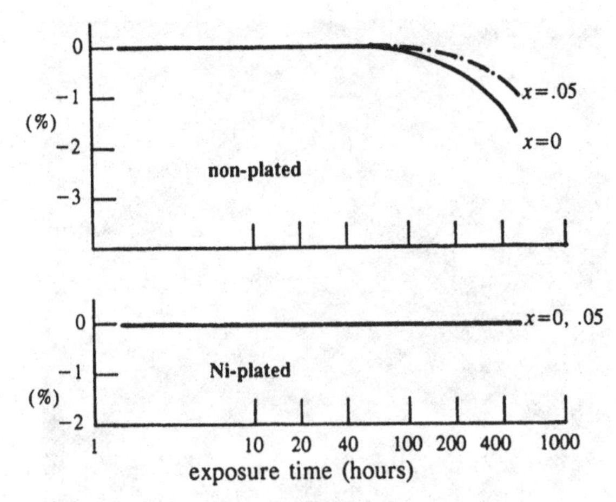

Figure 3.21. Flux loss in sintered $Nd_{15}(Fe_{1-x}Co_x)_{77}B_7Al_1$ $(x=0, 0.05)$ in 95% humidity at 60 °C (non-plated) and 80 °C (Ni-plated).

and at least three methods achieve this successfully. Figure 3.21 shows that a nickel-plated coating of 10–15μm thickness effectively stabilizes a typical $Nd_{15}(Fe_{1-x}Co_x)_{77}B_7Al_1$ magnet against flux loss in 95% humidity at 80 °C (Ohashi *et al.*, 1987). Other successful methods of impeding corrosive attack include epoxy-type coatings of 20–30μm thickness, and an aluminum layer about 10μm, when applied by ion vapor deposition to ensure good homogeneity. The inherent porosity of *bonded* magnets makes the process step of blending the compound particularly important, because at this stage the binder may be mixed in such a way as to coat the individual particles.

References

Adler, E. and Marik, H.-J. (1981). Stability of samarium–cobalt magnets. *5th International Workshop on Rare Earth–Cobalt Permanent Magnets and their Applications*, pp. 335–56. Dayton: University of Dayton.

Jacobson, J. and Kim, A. (1987). Oxidation behavior of Nd–Fe–B magnets. *Journal of Applied Physics*, **61**, 3763–5.

Jain, G. C. (1967). *Properties of Electrical Engineering Materials*. New York: Harper & Row.

Lee, R. W., Brewer, E. G. and Schaffel, N. A. (1985). Processing of neodymium–iron–boron melt-spun ribbons to fully dense magnets. *IEEE Transactions on Magnetics*, **21**, 1958–63.

Leupold, H. A., Potenziani, E., Clarke, J. P. and Tauber, A. (1984). High energy product, temperature compensated permanent magnets for device use at high operating temperatures. *IEEE Transactions on Magnetics*, **20**, 1572–4.

Narasimhan, K. S. V. L. (1981). Higher energy product rare earth–cobalt permanent magnets. *5th International Workshop on Rare Earth–Cobalt Permanent Magnets and their Applications*, pp. 629–55. Dayton: University of Dayton.
Ohashi, K., Tawara, Y., Yokoyama, T. and Kobayashi, N. (1987). Corrosion resistance of cobalt-containing Nd–Fe–Co–B magnets. *9th International Workshop on Rare-Earth Magnets and their Applications*, pp. 355–61. Bad Honnef: Deutsche Physikalische Gesellschaft.
Tenaud, P., Vial, F. and Sagawa, M. (1990). Improved corrosion and temperature behavior of modified Nd–Fe–B magnets. *IEEE Transactions on Magnetics*, **26**, 1930–2.
Tokunaga, M., Endoh, M. and Harada, H. (1987). Thermal stability of Nd–Fe–B sintered magnets. *9th International Workshop on Rare-Earth Magnets and their Applications*, pp. 477–84. Bad Honnef: Deutsche Physikalische Gesellschaft.
Wang, J. and Chang, K. (1987). Temperature compensated $Sm_{1-x}Gd_x(Co, Cu, Fe, Zr)_{7.4}$ permanent magnets. *9th International Workshop on Rare-Earth Magnets and their Applications*, pp. 411–14. Bad Honnef: Deutsche Physikalische Gesellschaft.
Willman, C. J. and Narasimhan, K. S. V. L. (1987). Corrosion characteristics of RE–Fe–B permanent magnets. *Journal of Applied Physics*, **61**, 3766–8.

4
Magnetic circuit design

4.1 Introduction

Analytical techniques, such as *finite element analysis*, provide accurate solutions for two- or three-dimensional field distributions in complex geometries, which in turn may be used to predict device performance with similar precision. However, these techniques require a detailed definition of the geometry and boundary conditions to be solved, which assumes that an initial design already exists. While providing an accurate field solution for a defined geometry, they will not optimize it – suggestions for changes to dimensions, materials, excitations, etc. must come from the designer, to be analyzed via a field solution. Consequently, while computer-based field analysis is an effective tool for simulating a known device, it is too cumbersome for design optimization.

Preliminary designs are usually performed using a magnetic circuit model of the device in which each component or magnetic flux path is represented by a discrete element. Equations representing the magnetic circuit components are derived in this Chapter, and the elements they define are used in an *equivalent* circuit (similar in many respects to an electrical analog circuit). This is a simple model, which can be easily optimized for any performance requirements. Thereafter, field analysis may be employed to verify the operation of the device, and to fine-tune the design.

We may consider that *flux* in a magnetic current is analogous to the *current* flowing in an electrical circuit, and we will demonstrate other similarities, which allow us to manipulate and solve an equivalent magnetic circuit in an identical manner to an electrical circuit. The difference, however, is that flux is not constrained as current is within electrical conductors, though the magnetic circuit designer strives to concentrate flux into those regions where it is most effective. The most difficult task is to define with sufficient accuracy the elements representing the surrounding regions, which carry non-effective or *leakage* flux, which in turn relies upon good assumptions of the paths followed by flux in these regions. The objective is to approximate the flux paths well enough to produce a preliminary design for the device, and to optimize its main parameters.

4.2 Field equations

Calculation of the flux in a magnetic circuit is based upon two very fundamental equations of electromagnetism, two of *Maxwell's Equations*, the first of which has already been derived in its differential form (Equation (1.27)) as *Ampère's Law*:

$$\boldsymbol{J} = \nabla \times \boldsymbol{H} \tag{1.27}$$

For magnetic circuit design, we need to manipulate this into its integral form, the first step being to integrate both sides over a surface element d$\boldsymbol{A}$:

$$\int \nabla \times \boldsymbol{H} \cdot \mathrm{d}\boldsymbol{A} = \int \boldsymbol{J} \cdot \mathrm{d}\boldsymbol{A} = i \tag{4.1}$$

Obviously the integral of current density $\boldsymbol{J}$ over an area is simply the current i flowing through that surface element.

When the *curl* or "$\nabla \times$" operator was introduced in Section 1.3, it was shown that it described *rotation* of the vector it operated upon. Consider that the element d$\boldsymbol{A}$ is part of a finite surface S spanning a loop L, as shown in Figure 4.1. If S is divided completely into elements such as d$\boldsymbol{A}$,

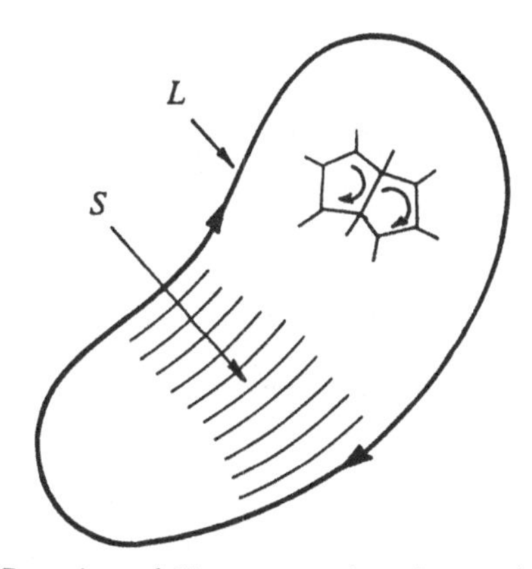

Figure 4.1. Rotation of H over a surface S spanning a loop L.

and the rotations $\boldsymbol{H}$ around all the elements are added together, then the contributions from internal boundary lines will cancel out because each such boundary is traversed first in one direction and then in the opposite direction in determining the contribution from an adjacent element. The only places where these contributions survive are along the outer boundary, so the result of this summation is the line integral of $\boldsymbol{H}$ around the loop L (Shercliff, 1977). This is a general result known as *Stokes' Theorem*:

$$\int \nabla \times \boldsymbol{H} \cdot \mathrm{d}A = \oint \boldsymbol{H} \cdot \mathrm{d}l \tag{4.2}$$

Combining Equations (4.1) and (4.2) shows that the closed line integral of $\boldsymbol{H}$ equals the total current that is linked by this closed path:

$$\oint \boldsymbol{H} \cdot \mathrm{d}l = i \tag{4.3}$$

The second fundamental relationship required for magnetic circuit design expresses mathematically the notion that flux must be conserved through any closed volume, i.e. it is continuous (violation of this rule implies the existence of a magnetic *monopole* within the volume). Consider the volume element $\delta x \cdot \delta y \cdot \delta z$ in Figure 4.2, for which inward flux densities B_x, B_y and B_z increase to outward values $B_x + \delta B_x$, $B_y + \delta B_y$ and $B_z + \delta B_z$ respectively. The net flux generated is found by multiplying each density

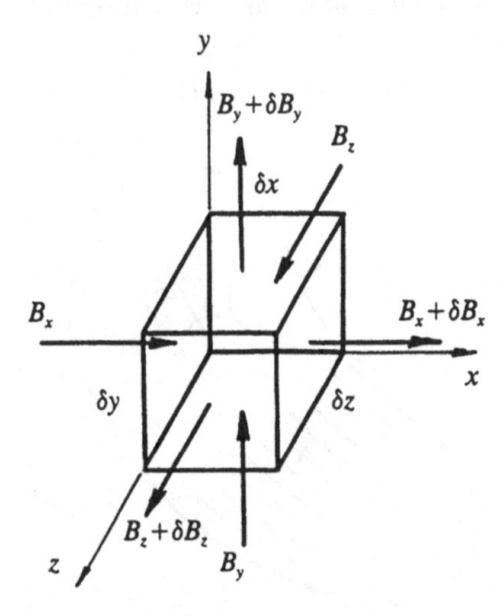

Figure 4.2. Flux conservation in volume element $\delta x\ \delta y\ \delta z$.

by the face area it crosses, and subtracting inward from outward flux. Flux conservation for this closed volume requires that this total be zero, so

$$\delta B_x\,\delta y\,\delta z+\delta B_y\,\delta z\,\delta x+\delta B_z\,\delta x\,\delta y=0 \tag{4.4}$$

This is simply the integral of flux density $\boldsymbol{B}$ through the volume element, performed over the surface enclosing that volume, and the shorthand expression for Equation (4.4) is

$$\oint \boldsymbol{B}\cdot \mathrm{d}\boldsymbol{A}=0 \tag{4.5}$$

Although it is not needed at this stage, it will be manipulated into its differential form for completeness. Rewriting Equation (4.4) per unit volume, and changing to partial differentials for the elemental volume,

$$\frac{\partial B_x}{\partial x}+\frac{\partial B_y}{\partial y}+\frac{\partial B_z}{\partial z}=0 \tag{4.6}$$

The three terms in this equation are components of $\boldsymbol{B}$ in the x, y and z directions respectively, and there is a shorthand notation for the operation that is performed here on a vector such as $\boldsymbol{B}$. It is the *div* operation, meaning *divergence*, which is written mathematically as "$\nabla\cdot$" (Shercliff, 1977). Equation (4.6) is therefore rewritten as

$$\nabla\cdot\boldsymbol{B}=0 \tag{4.7}$$

4.3 Static operation

The two basic relationships used for magnetic circuit design are Equations (4.3) and (4.5). Integrating the *magnetizing force* $\boldsymbol{H}$ around a closed loop yields the current i through the surface spanning that loop. In the case of a permanent magnet alone, surrounded only by air, there are no such currents and $i=0$ reduces Equation (4.3) to

$$\oint \boldsymbol{H}\cdot \mathrm{d}\boldsymbol{l}=0 \tag{4.8}$$

Figure 1.13 demonstrated the conservation of $\boldsymbol{B}$ across a magnet/air boundary, and the *demagnetizing* force $-\boldsymbol{H}$ within the material. More generally, though, the vector nature of the relationship between the magnetic parameters will be evident, as described by

$$\boldsymbol{B}=\mu_0(\boldsymbol{H}+\boldsymbol{M}) \tag{1.26}$$

Figure 4.3 shows a typical situation at two neighboring points, one within and one outside a magnet and close to its edge. The magnetic flux is continuous across the boundary, but $\boldsymbol{B}$, $\boldsymbol{H}$ and $\boldsymbol{M}$ are not aligned here within the magnet.

Equation (4.8) may be applied to the path L shown in Figure 4.3, part of which is in the permanent magnet and part of which is in air. Splitting the integral into these two components,

$$\int_{\text{magnet}} \boldsymbol{H}\cdot d\boldsymbol{l} + \int_{\text{gap}} \boldsymbol{H}\cdot d\boldsymbol{l} = 0 \tag{4.9}$$

Because the magnet experiences a demagnetizing force, the negative sign belongs to its $\boldsymbol{H}$, and Equation (4.9) is rewritten

$$\int_{\text{gap}} \boldsymbol{H}\cdot d\boldsymbol{l} = \int_{\text{magnet}} -\boldsymbol{H}\cdot d\boldsymbol{l} \tag{4.10}$$

It appears from this that $-\boldsymbol{H}$ in the magnet causes $+\boldsymbol{H}$ in the air gap, although the equality also involves the path length in each region. The product $\int \boldsymbol{H}\cdot d\boldsymbol{l}$ is known as the *magnetomotive force*, or "*m.m.f.*", which, from Equation (4.3), has the unit ampère-turns. If a permanent magnet has inadequate magnetizing force, then this may be compensated by additional magnet length to yield the desired m.m.f.

Consider Equations (4.5) and (4.8) applied to the very simple magnetic circuit of Figure 4.4, again having no real currents i. We shall make a number of assumptions initially, to clarify the derivation of the design equations. Soft iron pole pieces are used to direct flux from the magnet into an air gap; these are assumed to be infinitely permeable, will therefore

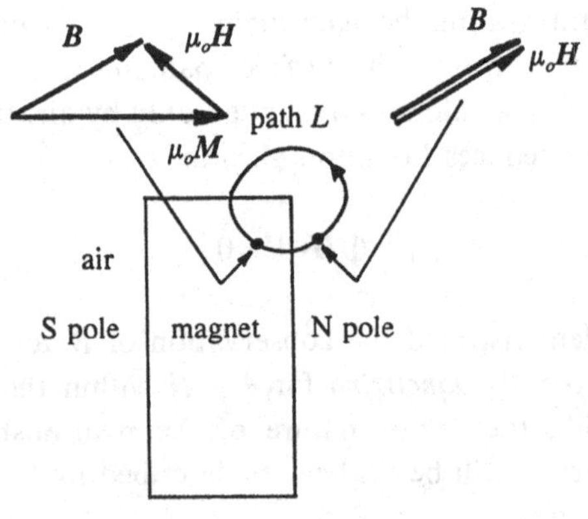

Figure 4.3. Field vectors either side of a magnet/air boundary, close to its edge.

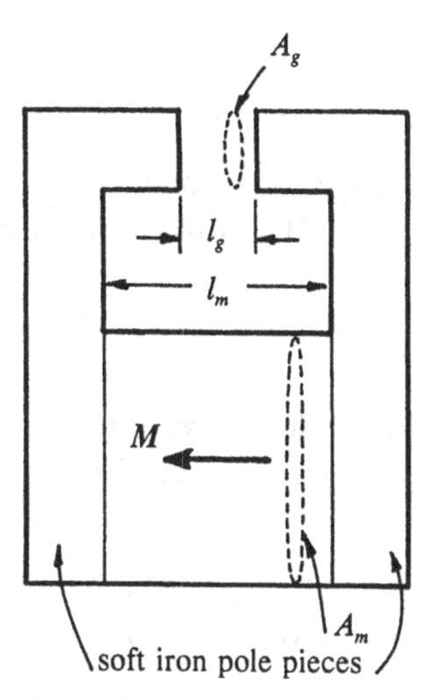

Figure 4.4. Magnetic circuit with a permanent magnet and an air gap.

require no magnetizing force ($\boldsymbol{H}=0$), and will permit no leakage flux. The length and cross-sectional area of the magnet (l_m, A_m) and the air gap (l_g, A_g) are well defined in this simplified example.

Technically, application of Equation (4.5) to this circuit requires that a closed surface be defined – consider this to pass down the middle of the circuit cutting through the air gap and the magnet, and closing anywhere outside the device. With no leakage flux, the total flux into (or out of) this surface will be

$$B_m A_m - B_g A_g = 0 \tag{4.11}$$

where B_m and B_g are flux densities in the magnet and air gap respectively. Either term represents the total flux flowing in the circuit, being

$$\Phi = B_m A_m = B_g A_g \tag{4.12}$$

All the flux generated by the magnet is delivered to the air gap.

Equation (4.8) may be applied to the circuit of Figure 4.4 around a closed path: along the axes of the magnet, each pole piece and the air gap. Since $\boldsymbol{H}=0$ in the pole pieces, Equation (4.8) once again breaks into its components in the manner of Equations (4.9) and (4.10):

$$H_m l_m + H_g l_g = 0 \tag{4.13}$$

$H_m l_m$ is the m.m.f. of the magnet, driving flux into the air gap, across which is an m.m.f. $H_g l_g$. Since $\boldsymbol{M}=0$ in the air gap, Equation (1.26) reduces to its familiar air gap form:

$$B_g = \mu_0 H_g \tag{4.14}$$

To determine the magnet's operating condition, B_g and H_g are eliminated between Equations (4.12), (4.13) and (4.14) to yield

$$B_m A_m = -\mu_0 A_g H_m \frac{l_m}{l_g} \tag{4.15}$$

Rearrangement of this provides a relationship between the magnet parameters B_m and H_m, which is a function of the magnetic circuit dimensions *only*:

$$\frac{B_m}{\mu_0 H_m} = -\frac{A_g l_m}{A_m l_g} \tag{4.16}$$

This is directly analogous to Equation (3.10), though much more convenient than the complex calculation of the demagnetizing factor N from the magnet's dimensions. Having the magnetic circuit geometry remain unchanged is one condition for *static* operation of the magnet, defined by Equation (4.16) as a straight line with a negative slope – known as the *load line* of the circuit. To determine the magnet's operating condition requires a second relationship between its two parameters, provided by its own demagnetization characteristic, which is modeled by Equation (1.26). As we have noted, the vector relationship between $\boldsymbol{B}$, $\boldsymbol{H}$ and $\boldsymbol{M}$ will actually vary from point to point within the material. In this simplified example, though, we will assume that a unique relationship exists along the axis of the magnet in the direction of the flux, written as

$$B_m = \mu_0 (H_m + M) \tag{4.17}$$

Any change in the state of magnetization M will cause the demagnetization characteristic to become non-linear, and a number of these curves for the most popular magnet materials were given throughout Chapter 3. The intercept between Equations (4.16) (the load line) and (4.17) (the demagnetization characteristic) occurs in the second quadrant of the latter, where H_m has a negative value; the magnet's operating point (B_m, H_m) is uniquely defined as shown in Figure 4.5. As was discussed in Chapter 3, the thermal stability of a magnet is improved by increasing the slope of the load line. According to Equation (4.16), this is achieved by adding length l_m to the magnet, or by reducing the air gap l_g. In the limit $l_g \rightarrow 0$, the magnet is short-circuited by a soft iron "keeper", and $B_m/\mu_0 H_m \rightarrow -\infty$ as shown in Figure 4.6. In practice, the upper limit to this slope is

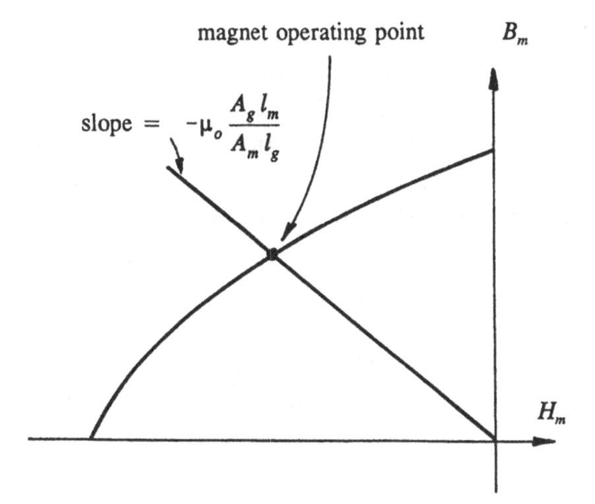

Figure 4.5. Demagnetization curve and load line for a magnet.

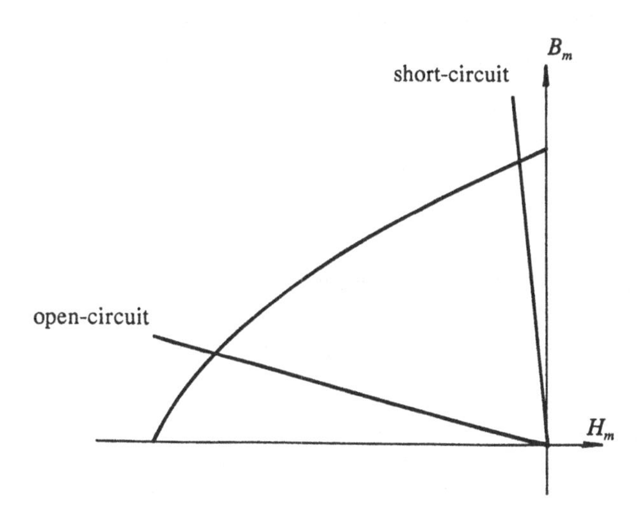

Figure 4.6. Demagnetization curve with short-circuit and open-circuit load lines for a magnet.

determined by a finite minimum gap at the surface contacts. Conversely, if the soft iron pole pieces are removed, the magnet stands alone and the air gap l_g becomes very large. This open-circuit condition is represented as shown in Figure 4.6 by a small load line slope.

Representation of an actual magnetic circuit will be more complex than this – the soft iron pole pieces have finite permeability, and there will be measurable leakage flux. Both these factors reduce the performance of the magnetic circuit, for which allowance should be made in the design equations. These effects are illustrated by defining correction factors k_1, to allow for the fact that not all the magnet flux reaches the air gap as "useful flux", and k_2, to account for magnetization of the (soft iron) circuit components connecting the magnet to the air gap. k_1 is sometimes known as the *leakage coefficient*, while k_2 is called the *loss factor*.

$$k_1 = \frac{\text{magnet flux}}{\text{useful flux}}$$

$$= \frac{\text{useful flux} - \text{leakage flux}}{\text{useful flux}} > 1 \tag{4.18}$$

$$k_2 = \frac{\text{magnet m.m.f.}}{\text{useful m.m.f.}} > 1 \tag{4.19}$$

Determining accurate values for k_1 and k_2 can be a most complex task, but it is essential to defining a realistic model of the magnetic circuit and a reliable prediction of device performance. The effect of these two factors upon the magnet's operating condition is illustrated by modifying Equations (4.12) and (4.13) to

$$B_{\mathrm{m}} A_{\mathrm{m}} = k_1 B_{\mathrm{g}} A_{\mathrm{g}} \tag{4.20}$$

$$H_{\mathrm{m}} l_{\mathrm{m}} + k_2 H_{\mathrm{g}} l_{\mathrm{g}} = 0 \tag{4.21}$$

The *load line* equation therefore becomes

$$\frac{B_{\mathrm{m}}}{\mu_0 H_{\mathrm{m}}} = -\left(\frac{k_1}{k_2}\right)\left(\frac{A_{\mathrm{g}} l_{\mathrm{m}}}{A_{\mathrm{m}} l_{\mathrm{g}}}\right) = -S \tag{4.22}$$

According to Equation (4.20), the reduction of useful flux due to leakage (k_1) can be compensated by increasing the magnet area (A_{m}), which also maintains the magnet's operating point. Likewise, Equation (4.21) shows that, if there is a measurable drop in m.m.f. through the soft iron components (k_2), then this may be compensated by raising the magnet's m.m.f. via its length (l_{m}).

4.4 Dynamic operation

If the magnetic circuit geometry does not remain unaltered, then the load line slope in Equation (4.22) will change via any of l_{g}, A_{g}, k_1 or k_2. The magnet's operating point may undergo an excursion within the short- and open-circuit limits shown in Figure 4.6. Another common way for a magnet to experience such *dynamic* operation is by direct excitation from a coil. In this case, the reduction of Equation (4.3) to (4.8) is invalid because i is now present.

Consider again the simple magnetic circuit of Figure 4.4, but with the addition of a coil of N turns carrying current i, as shown in Figure 4.7. When Equation (4.3) is applied to this circuit around the same closed path, the surface spanning this loop links the current i a total of N times. The subsequent equations involving m.m.f. must be modified to include this term, and Equation (4.21) becomes

$$H_m l_m + k_2 H_g l_g = Ni \tag{4.23}$$

This emphasizes the units of the m.m.f. equations being ampère-turns. Combining Equations (4.14), (4.20) and (4.23) provides a modified *load line*, which includes coil excitation:

$$B_m = -\mu_0 \left(\frac{k_1}{k_2}\right)\left(\frac{A_g l_m}{A_m l_g}\right)\left(H_m - \frac{Ni}{l_m}\right) \tag{4.24}$$

The *slope* of the load line is still determined only by the circuit dimensions and constants, Equation (4.24) reducing to its prior form as Equation (4.22) when $i=0$.

The *position* of the load line, however, is determined by the coil excitation, the intercept of the load line with the H_m axis being displaced by $(Ni)/l_m$ as shown in Figure 4.8. The direction of i shown in Figure 4.8

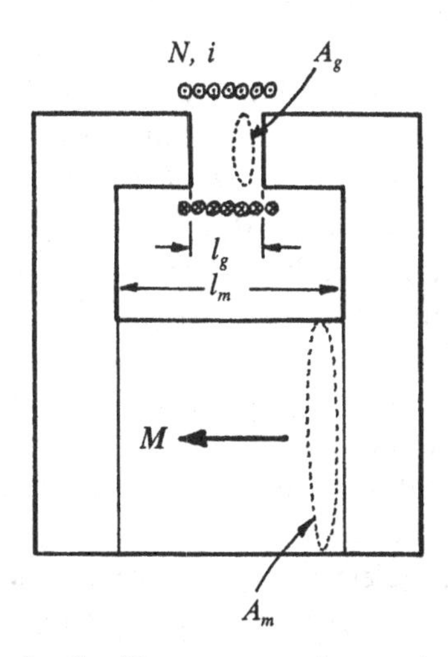

Figure 4.7. Magnetic circuit with a permanent magnet, an air gap and a coil.

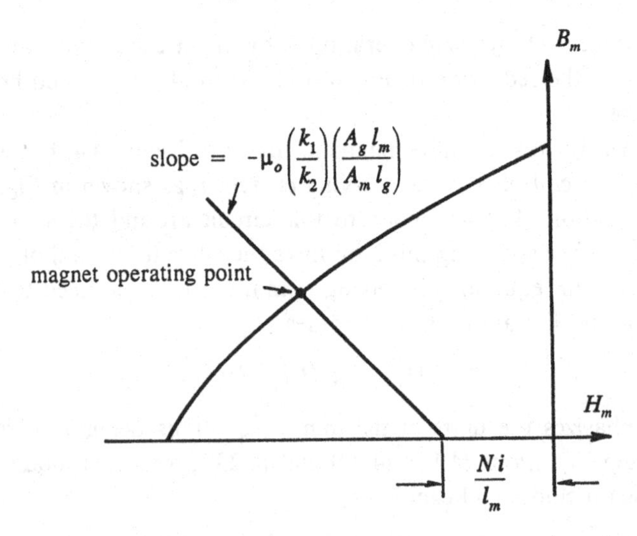

Figure 4.8. Demagnetization curve and load line for a magnet under coil excitation.

is consistent with the integration path, reducing the magnitude of H_m in accordance with Equation (4.24), and enhancing the flux in the circuit. Conversely, if i flows opposite to the direction shown in Figure 4.7, it increases $|H_m|$ moving the load line to the left as in Figure 4.8, reducing the magnitude of the flux. The amplitude of these lateral excursions of the magnet's operating point via its magnetization is also dependent upon l_m, being greater the shorter the magnet. Therefore, while additional magnet length may be used to raise the slope of the load line, this will also stabilize the operating point against externally applied fields.

For an ideal magnet based upon magnetocrystalline anisotropy, it has been shown with reference to Figure 1.11 that, with $|H_{ci}| > M_{sat}$, the second quadrant demagnetization curve is entirely linear, corresponding to which the magnetization is the constant M_{sat}. With shape anisotropy, however, $|H_{ci}| \leqslant M_{sat}$ and there is a knee due to $-H_{ci}$ in the demagnetization curve, as was shown in Figure 1.17. The characteristics of practical permanent magnets are not as linear as these ideals suggest, and, as discussed in Chapter 3, they will continuously vary with temperature. Figure 3.4 showed that if the operating point falls below the knee of the demagnetization curve, an *irreversible* change takes place – operation on the "major" curve can only be restored by full remagnetization of the material. Obviously the same problem can occur with a sufficient decrease in load line slope (usually due to opening up the air gap) and/or with a shift in load line position at high enough applied demagnetizing field.

The definition of *intrinsic coercivity* is the value of applied field required to reverse the saturation magnetization, whether as $-H_{ci}$ that the magnet can withstand before passing the knee, or as $+H_{ci}$ needed to fully magnetize the material. From any point beyond the knee of the demagnetization curve, such as those shown in Figure 4.9, $+H_{ci}$ will be required to reverse those domains now having $-M_{sat}$ back to $+M_{sat}$. The magnet therefore follows the path of a "minor" *B versus H* curve until it rejoins the "major" loop in the first quadrant. Full remagnetization will not be achieved unless a field of $+H_{ci}$ at least is achieved.

It may not be practical to fully remagnetize a material that has suffered an *irreversible* loss beyond the knee, for in many cases the magnet will continue to experience that same, repeated change in load line slope or position. Consider, for example, the magnet operating initially above the knee at point ***a*** in Figure 4.10. A change in load line slope or position then shifts the magnet to operating point ***b***, below the knee of the curve. This change is irreversible, so even when it is then removed, the magnet returns not to ***a*** but to operating point ***c*** along a "minor" *B versus H* curve. Subsequent application and removal of the same load line disturbance will cause the magnet to operate *reversibly* along this minor curve between points ***b*** and ***c***.

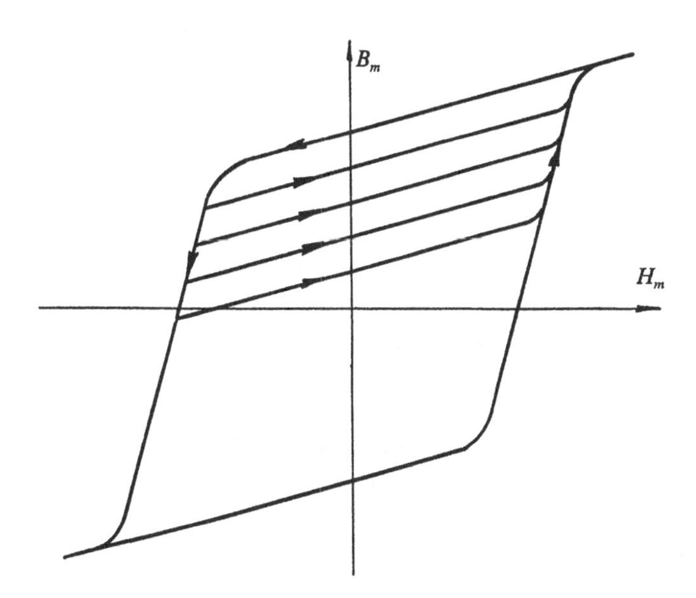

Figure 4.9. Remagnetization of a magnet.

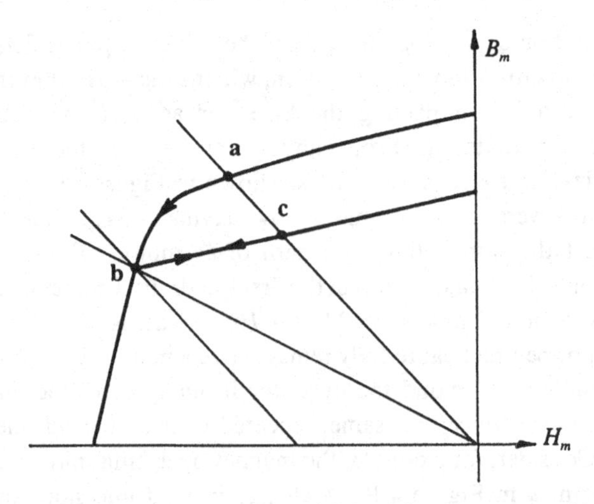

Figure 4.10. Change in operation with load line slope or position.

Dynamic operation of a magnet in normal service results in movement of its operating point (B_m, H_m) either along the major or on minor B *versus* H curves. When caused by changes in geometry, this movement is bounded within the second quadrant as shown in Figure 4.6. When caused by applied excitation, though, the operating point may move into the third quadrant (where B_m is reversed), or into the first quadrant (though this is uncommon in practical magnet applications). The portions of the minor curves that lie in the second (or third) quadrant are also known as *recoil lines*, and dynamic magnet operation such as that between points ***b*** and ***c*** in Figure 4.10 is usually known as *recoil*. Because recoil lines are the initial segments of minor curves, they are essentially *linear* and of the same *slope*, regardless of their origin on the major B *versus* H loop. This is a useful approximation for design purposes, though measurements reveal that the *recoil lines* are actually very narrow *loops*, as shown in Figure 4.11 for Alnico 5. The *slope* of the recoil lines is a fundamental characteristic of a permanent magnet, given as its relative *recoil permeability*, μ_{rec}.

Many permanent magnets, especially the ceramic ferrite and rare earth materials, have demagnetization characteristics that are linear throughout much or all of the second quadrant, thanks to their high H_{ci}. As explained in the derivation of the B *versus* H characteristic in Figure 1.11 from its intrinsic counterpart in Figure 1.10, this linearity results from an essentially constant value of $\boldsymbol{M} = +M_{sat}$ up to $-H_{ci}$. This is of particular benefit in dynamic operation, because if the operating point is constrained to this linear region of the characteristic, then recoil cannot occur *within* the

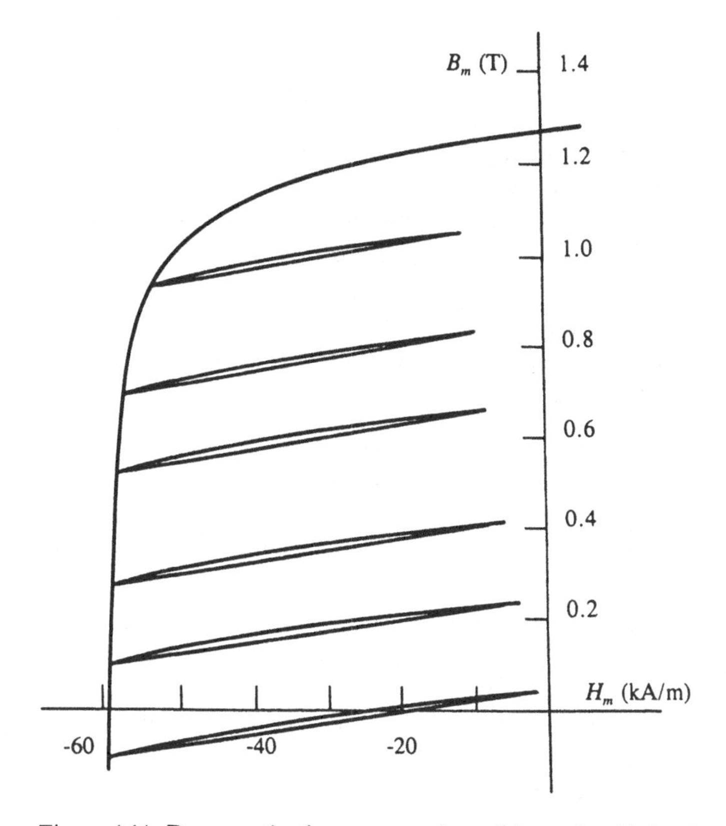

Figure 4.11. Demagnetization curve and recoil loops for Alnico 5.

major B *versus* H loop on a minor curve. This is illustrated by differentiating Equation (4.17), which models the demagnetization curve:

$$\frac{\mathrm{d}B_\mathrm{m}}{\mathrm{d}H_\mathrm{m}} = \mu_0\left(1 + \frac{\mathrm{d}M}{\mathrm{d}H_\mathrm{m}}\right) \tag{4.25}$$

Over the range where M is *constant*, this slope is also of constant μ_0. Recoil represents an irreversible loss of magnet energy, and occurs within the major B *versus* H loop – the slope of the recoil line can be no greater than that on the major curve where it originated, but no less than μ_0, the permeability of free space (air). Therefore, $\mu_{\mathrm{rec}} = 1$ and dynamic operation of the magnet is constrained to a linear demagnetization characteristic. If M is *not constant*, though, it will have values in the second quadrant that are less than M_{sat}, and $\mu_{\mathrm{rec}} > 1$ is allowed – a recoil line is represented by $\mu_0\mu_{\mathrm{rec}}$ and $\mu_0 M_\mathrm{r}$, its intercept on the B_m axis ($< \mu_0 M$ of the major B

versus H curve), as

$$B_m = \mu_0(\mu_{rec}H_m + M_r) \tag{4.26}$$

In a magnetic circuit containing both a permanent magnet and a coil, it is sometimes necessary to calculate the effect of coil excitation alone, as in the determination of its inductance. The foregoing analysis indicates that the magnet appears to the coil simply as its recoil permeability $\mu_0\mu_{rec}$. If the magnet is operating on a linear demagnetization characteristic, then it will appear as a pure air gap.

An operating load line for a magnet was previously derived in terms of its *internal* field as

$$\frac{\boldsymbol{B}_i}{\mu_0 \boldsymbol{H}_i} = \frac{\boldsymbol{M}(1-N)}{H_o - N\boldsymbol{M}} \tag{3.10}$$

This slope will clearly change if the magnetization $\boldsymbol{M}$ is *not* constant, which is certainly the case when the operating point approaches and passes beyond the knee of the "major" loop. Flux density B_m is not the fundamental property of a magnet, but is a consequence of $\boldsymbol{M}$ via Equations (1.26) and (4.17). Under high demagnetizing fields, which cause the magnet to operate around its knee, it is common to use the intrinsic characteristic to determine the operating condition, and then to subsequently deduce B_m. Equation (4.17) is used with (4.22) to derive a *load line* that is appropriate for the intrinsic curve as

$$\frac{M}{H_m} = -(S+1) \tag{4.27}$$

Likewise, Equations (4.17) and (4.24) are combined to include the effect of coil excitation upon the load line:

$$M = -(S+1)H_m + S\left(\frac{Ni}{l_m}\right) \tag{4.28}$$

The slope of the load line, which is determined only by the magnetic circuit topology, is $-S$ on a B *versus* H diagram and $-(S+1)$ on the intrinsic characteristic. Furthermore, the lateral excursion of each load line due to coil excitation is different in each case, Ni/l_m in Figure 4.8 being scaled by the factor $S/(S+1)$ as shown in Figure 4.12. After finding the magnetizing force H_m from the intrinsic curve, a vertical line as in Figure 4.12 may be used to locate the value of B_m.

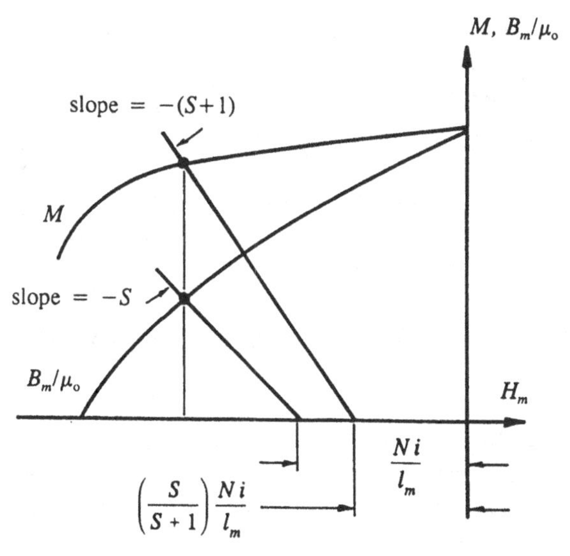

Figure 4.12. Intrinsic demagnetization characteristic and load line for a magnet under coil excitation.

4.5 Energy transfer

The foregoing has presented the method of determining a magnet's operating point (B_m, H_m), associated with which is a stored energy, which may be instrumental in the conversion of electrical and mechanical work. Section 1.5 presented a derivation of the energy stored in volume V of a magnet, leading to the change in energy from point ***a*** to point ***b*** on the *B versus H* loop shown in Figure 1.12:

$$-\int_a^b B\,\mathrm{d}H - \int_a^b H\,\mathrm{d}B = -[BH]_a^b \tag{1.48}$$

Per unit volume of the magnet material, the first term is the work done by the applied field and the second is the internal kinetic energy stored, the sum of which equals the total potential energy.

After being magnetized to saturation, the potential energy is reduced to zero at remanence ($B_m = 0$ at point ***b***); the magnet must move into the second quadrant to deliver its stored energy. Figure 4.13 shows how the three energy density components of Equation (1.48) develop as the magnet moves from point ***b*** to point ***c***. Considering the direction of integration, the three areas sum according to Equation (1.48), the work done as applied field $\int B\,\mathrm{d}H$ equalling the change in potential energy $[BH]$ plus the release

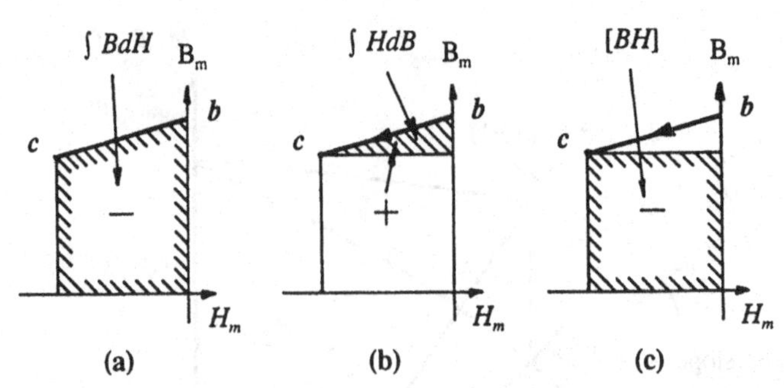

Figure 4.13. Change per unit volume of (*a*) applied field energy, (*b*) magnet kinetic energy and (*c*) total potential energy.

of kinetic energy from the magnet $\int H\,dB$. The higher the potential energy of a permanent magnet, the greater will be the release of its kinetic energy in establishing an external field. While modern magnet developments have focussed on improving the available (potential) energy density, it clearly requires a corresponding increase in the work to change the operating condition of the material. Consequently, the basis of most devices using high energy magnets is a design that keeps the magnet close to a unique operating point with a minimum of dynamic operation.

The kinetic energy released by a magnet of volume V, which is operating at a typical point c in Figure 4.13, is

$$E = V\int H\,dB \tag{4.29}$$

This general result may be applied to the field in any volume, even an air gap V_g represented by Equation (4.14):

$$\begin{aligned} E_g &= V_g \int \frac{B_g}{\mu_0}\,dB_g \\ &= \frac{V_g B_g^2}{2\mu_0} \\ &= \frac{V_g B_g H_g}{2} \end{aligned} \tag{4.30}$$

If this is the air gap in the simple magnetic circuit of Figure 4.4, then Equations (4.11) and (4.13) may be used in (4.30) to demonstrate that all the energy released from the magnet is delivered into the gap:

$$E_g = \frac{V_g B_g H_g}{2} = \frac{-V_m B_m H_m}{2} \tag{4.31}$$

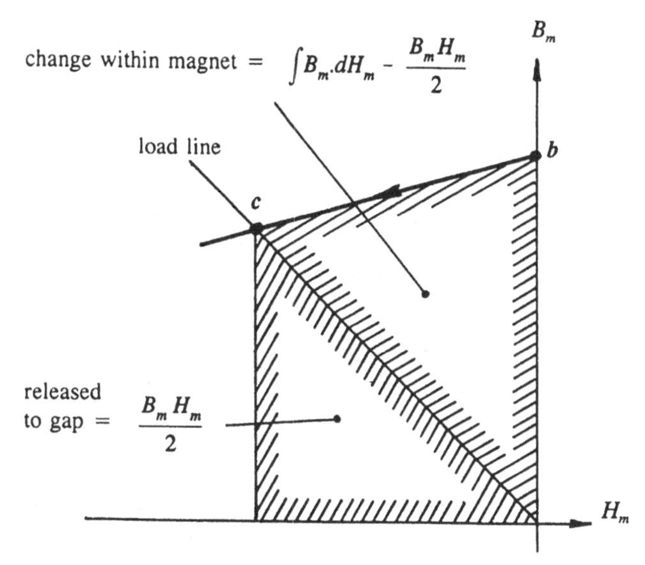

Figure 4.14. Change per unit volume of magnet's applied field energy.

Notice that the *energy density* delivered to the gap is *not* the magnet's *energy product*, but *half* of this value. Neither is it the work done by the magnet's applied field associated with it establishing E_g – redrawing Figure 4.13(*a*) as Figure 4.14 shows that $\int B\,\mathrm{d}H$ actually has two components. One is the energy per unit *magnet* volume released into the gap, according to Equation (4.31). The remainder is the area swept by the *load line* in establishing this operating condition for the magnet, the change in the applied field energy within the magnet as a gap l_g is opened from ***b*** to ***c***.

The further down its demagnetization curve a magnet is to be driven, the greater is the area swept by the load line, for which the applied field within the magnet does more work. This energy is recaptured by the magnet if the gap is re-closed and the load line returns to its original position. However, if this cycle includes driving the magnet beyond a knee in its characteristic, a return to the original load line along a recoil line involves the *irreversible* loss of magnet field energy shown in Figure 4.15. Again, the desirability of minimizing the dynamic operation of a high energy magnet is demonstrated – significant excursions should be restricted if possible to increasing l_g during installation of the magnet. Subsequent cycling along the recoil line incurs no further irreversible loss, only reversible changes in energy.

While $(BH)_{max}$ is the most common figure of merit for a permanent magnet, operation at this point maximizes the release of magnet energy

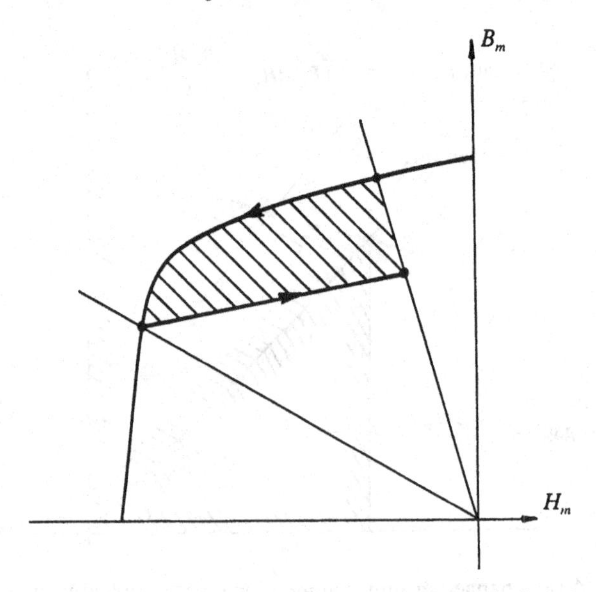

Figure 4.15. Change in magnet field energy with load line slope.

into the gap at $-\frac{1}{2}V_m(BH)_{max}$. If the magnetization M is constant and the linear demagnetization characteristic is represented by Equation (4.17), then the air gap energy (Equation (4.31)) may be written as

$$E_g = \frac{-\mu_0 V_m}{2}(H_m^2 + MH_m) \tag{4.32}$$

Differentiation of this finds that $(BH)_{max}$ occurs at $B_m = \frac{1}{2}\mu_0 M$, $-H_m = \frac{1}{2}M$, for which

$$(E_g)_{max} = \frac{\mu_0 V_m}{2}\left(\frac{M}{2}\right)^2 \tag{4.33}$$

The magnet energy released is smaller either side of the $(BH)_{max}$ point on the major demagnetization characteristic, and by virtue of a reduced alignment of the magnetization M, E_g is also smaller for operation on recoil lines within the major B *versus* H curve. It is helpful to superimpose constant energy contours on the characteristic to demonstrate the penalty of such operation, though the normal convention is to express these as constant *energy product* $(B_m H_m)$ as shown in Figure 4.16, rather than the true *energy density* $(\frac{1}{2}B_m H_m)$.

A more realistic representation of the soft iron pole pieces requires inclusion of the flux leakage coefficient k_1 and m.m.f. loss factor k_2 via

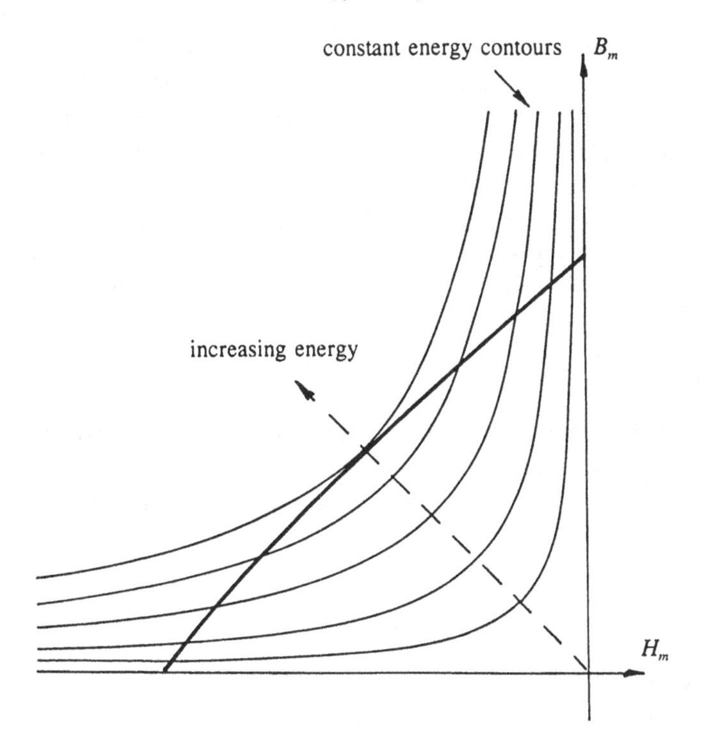

Figure 4.16. Demagnetization characteristic and constant energy contours.

Equations (4.20) and (4.21) with (4.30). As would be expected, both these factors reduce the amount of magnet energy that is delivered to the gap:

$$E_g = \frac{V_g B_g H_g}{2} = \frac{-V_m B_m H_m}{2k_1 k_2} \tag{4.34}$$

Next, add the coil of N turns with current i, and using Equation (4.23) rather than (4.21), the air gap energy becomes

$$E_g = \frac{-V_m B_m}{2k_1 k_2}\left(H_m - \frac{Ni}{l_m}\right) \tag{4.35}$$

It is informative to rewrite the second term representing the energy contribution from the coil, noting that the flux in the circuit is $\Phi = B_m A_m$, and the total flux *linkage* with the coil is defined as $\lambda = N\Phi$:

$$E_g = \frac{-V_m B_m H_m}{2k_1 k_2} + \frac{\lambda i}{2k_1 k_2} \tag{4.36}$$

While a permanent magnet usually establishes a relatively constant field in a magnetic circuit, the contribution from a coil is frequently the cause or result of electromechanical energy conversion. The appearance of λi in the coil term of Equation (4.36) is not surprising, and can be derived from Equation (4.29), the general expression for energy in any magnetic component. For a coil alone, m.m.f. $Ni = Hl$, flux $\Phi = B/A$, and $\lambda = N\Phi$, the corresponding general expression for energy released from a coil of area A and length l will be

$$E = \int i \, \mathrm{d}\lambda \tag{4.37}$$

This would be utilized for a coil in a similar manner to the way Equation (4.29) was employed for a permanent magnet. It also shows that the current induced in a coil may be determined by differentiating its energy with respect to λ:

$$i = \frac{\mathrm{d}E}{\mathrm{d}\lambda} \tag{4.38}$$

Rate of change of energy is the electrical *power* in the component:

$$\frac{\mathrm{d}E}{\mathrm{d}t} = i\frac{\mathrm{d}\lambda}{\mathrm{d}t} = -ei \tag{4.39}$$

The induced e.m.f. $-e$ in a coil is defined by *Faraday's Law* of electromagnetic induction as the rate of change of its flux linkage. Electromechanical energy conversion is the transformation of electrical power into mechanical power, or vice versa. Adding the mechanical parameters force F and velocity v to Equation (4.39) yields (Woodson and Melcher, 1968):

$$\frac{\mathrm{d}E}{\mathrm{d}t} = -ei - Fv \tag{4.40}$$

Mechanical power is defined here as acting *into* the component, so a minus sign is also associated with F. Since $v = \mathrm{d}x/\mathrm{d}t$, the change in energy is

$$\mathrm{d}E = i \, \mathrm{d}\lambda - F \, \mathrm{d}x \tag{4.41}$$

The conversion between electrical and mechanical energy is perfect if $\mathrm{d}E = 0$, though fortunately we are able to evaluate these two terms *independently*, treating $\mathrm{d}\lambda$ and $\mathrm{d}x$ as consecutive events. Although a device may change from state 1 to state 2 by some complex route as shown in

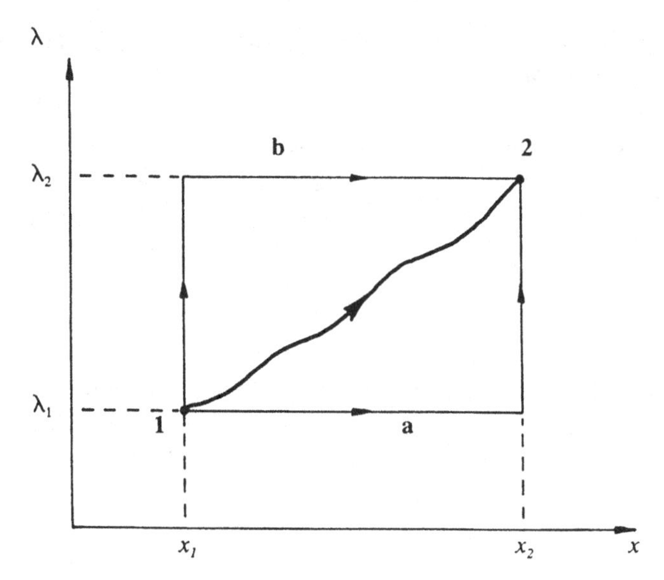

Figure 4.17. Change in energy state using independent variables.

Figure 4.17, this may be treated as a change of dimension x followed by a change in flux linkage λ (path a), or vice versa (path b). Expressed mathematically, the energy difference between the two states via paths a and b is described by Equations (4.42) and (4.43) respectively:

$$\Delta E = -\int_{x_1}^{x_2} F(\lambda_1, x)\,\mathrm{d}x + \int_{\lambda_1}^{\lambda_2} i(\lambda, x_2)\,\mathrm{d}\lambda \tag{4.42}$$

$$\Delta E = \int_{\lambda_1}^{\lambda_2} i(\lambda, x_1)\,\mathrm{d}\lambda - \int_{x_1}^{x_2} F(\lambda_2, x)\,\mathrm{d}x \tag{4.43}$$

Treating λ and x as independent variables allows i still to be found by differentiating the energy with respect to λ, while mechanical force (in the direction of x) may be determined by differentiation with respect to x:

$$F = -\frac{\mathrm{d}E}{\mathrm{d}x} \tag{4.44}$$

In the more common case of a rotating device, the torque T is found with respect to angular position θ via

$$T = -\frac{\mathrm{d}E}{\mathrm{d}\theta} \tag{4.45}$$

This relationship was used in Equation (1.6) when we evaluated the torque on a magnetic dipole current loop.

As a simple example of one of these relationships, we shall use Equation (4.44) to calculate the force of attraction between the two pole faces bounding the air gap of Figure 4.4 (Figure 4.18) of area A_g and length l_g. The energy in this gap is given by Equation (4.30), which, with $B_g = \mu_0 H_g$, becomes

$$E_g = \frac{A_g l_g B_g^2}{2\mu_0} \tag{4.46}$$

The air gap dimensions contain the three Cartesian axes in which force may be calculated – in this case, we want to define the x direction in Equation (4.44) across the gap, along its length l_g. Differentiating Equation (4.46) with respect to l_g yields

$$\begin{aligned} F_x &= -\frac{\mathrm{d}E_g}{\mathrm{d}l_g} \\ &= -\frac{A_g B_g^2}{2\mu_0} = -\frac{\Phi^2}{2\mu_0 A_g} \end{aligned} \tag{4.47}$$

Our definition of $-F_x$ acting *into* the air gap confirms this as a force of *attraction*, which will increase as the gap closes; while l_g is not contained in Equation (4.47), F_x is proportional to Φ^2, the flux rising as a decreasing

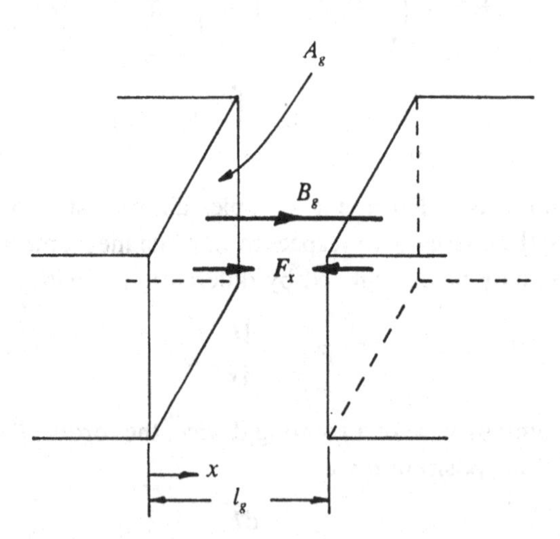

Figure 4.18. Force of attraction between opposite pole faces across an air gap.

l_g causes the load line slope to raise the magnet's operating point up its demagnetization characteristic. Similarly, there are forces acting to retard the lateral displacement of one pole piece with respect to the other (F_y, F_z), which may also be determined from Equation (4.46) by splitting A_g into its y and z components and differentiating with respect to the appropriate axis.

4.6 Equivalent magnetic circuit

The examples used thus far were simple magnetic circuits comprising a permanent magnet, a single air gap, and perhaps a coil – real magnetic circuits are generally more complex than this. For example, the effect of leakage flux was accounted for by the correction factor k_1, but this is an important effect, which normally requires calculation using good estimates of leakage gap areas and lengths. By its nature, leakage flux follows paths in parallel with the main air gap in a magnetic circuit, which complicates the calculation of load line slope – A_g and l_g must represent *net* load as seen by the magnet. While it seems convenient to find the operating point of a magnet by the intersection of its load line with its demagnetization characteristic, computing the net load line for a complex magnetic circuit may not be that easy. A simpler approach involves defining a characteristic parameter for each circuit component – its *permeance*.

A model of the magnetic circuit will be developed in which each component is represented by its dimensions and material properties. This will not involve use of correction factors such as k_1 and k_2, so a load line such as Equation (4.24) may be rewritten as

$$B_m A_m = -\mu_0 \frac{A_g}{l_g} (H_m l_m - Ni) \tag{4.48}$$

This is simply an expression of flux Φ_m in the magnet, in terms of the m.m.f.s of the magnet F_m $(=H_m l_m)$ and the coil Ni:

$$\Phi_m = -\mu_0 \frac{A_g}{l_g} (F_m - Ni) \tag{4.49}$$

Unlike Equation (4.24), this linear relationship is a *load line* whose slope contains the parameters of one circuit component only – the air gap load. By defining the m.m.f. across the gap as $F_g = H_g l_g$, Equation (4.14) becomes

$$\Phi_g = \mu_0 \frac{A_g}{l_g} F_g = P_g F_g \tag{4.50}$$

where the *permeance* of the gap is defined by its parameters as

$$P_g = \frac{\mu_0 A_g}{l_g} \tag{4.51}$$

This is the slope of a flux *versus* m.m.f. load line related either to this air gap or to the magnet:

$$\Phi_m = -P_g(F_m - Ni) \tag{4.52}$$

The same transformation is performed on Equation (4.17) for the demagnetization characteristic to find the magnet's operating point. The magnet's permeance is likewise defined by its own area and length as

$$P_m = \frac{\mu_0 A_m}{l_m} \tag{4.53}$$

Hence

$$\Phi_m = P_m(F_m + Ml_m) \tag{4.54}$$

Equations (4.52) and (4.54) are plotted in Figure 4.19, which is equivalent to Figure 4.8. The original demagnetization characteristic is easily scaled

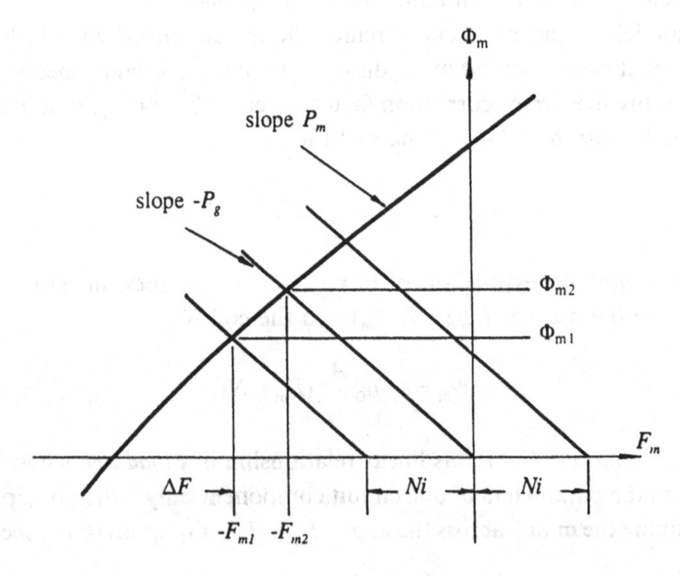

Figure 4.19. Demagnetization curve and load line for a magnet expressed as permeances.

into this new form, the coercivity being multiplied by magnet length to give the equivalent m.m.f. $H_c l_m$. A convenient feature now is that the lateral displacement of the new load line is by the actual coil excitation Ni. A coil current $-i$ as shown moves the magnet's operating point from $-F_{m2}$, Φ_{m2} to $-F_{m1}$, Φ_{m1}, and the following relationships are deduced from the geometry of Figure 4.19:

$$P_m = \frac{\Phi_{m2} - \Phi_{m1}}{F_{m1} - F_{m2}}$$

$$-P_g = -\frac{\Phi_{m2}}{F_{m2}} = -\frac{\Phi_{m1}}{\Delta F} \qquad (4.55)$$

$$Ni = F_1 - \Delta F$$

The change in magnet flux due to a coil excitation Ni is found by eliminating between the Equations (4.55):

$$\Delta\Phi_m = \Phi_{m2} - \Phi_{m1} = \frac{Ni}{P_m^{-1} + P_g^{-1}} \qquad (4.56)$$

$\Delta\Phi_m$ can be reduced in a magnetic design by decreasing either of the slopes P_m or P_g, which might be achieved by raising l_m or l_g respectively – additional magnet length has the effect of stabilizing the change of flux in the magnet.

In some applications, it is desired to *maximize* the change in flux for a given excitation, which requires increasing P_m and/or P_g. This is usually achieved not by reducing l_m or l_g, but by raising the permeability of the component materials. The magnet should be operating on a recoil line of slope $\mu_0\mu_{rec}$, which was represented by Equation (4.26). Changing the definition of P_m to include μ_{rec}, Equations (4.53) and (4.54) become

$$P_m = \frac{\mu_0 \mu_{rec} A_m}{l_m} \qquad (4.57)$$

$$\Phi_m = P_m\left(F_m + \frac{M l_m}{\mu_{rec}}\right) \qquad (4.58)$$

To improve $\Delta\Phi_m$ (Equation (4.56)) one might therefore choose to employ a magnet with a *high* recoil permeability μ_{rec}, such as one of the alnico materials. The same is true for the "air" gap, which one might consider to be a soft magnetic material of relative permeability μ. The load line

Equation (4.52) is unchanged provided that the component is now defined by

$$B_g = \mu_0 \mu H_g \tag{4.59}$$

$$P_g = \frac{\mu_0 \mu A_g}{l_g} \tag{4.60}$$

These are just more general versions of the equations for air, for which $\mu = 1$.

The analogy of an equivalent magnetic circuit to an electrical circuit is fairly obvious. Magnetic flux (c.f. electrical current) flows out of a magnet with an m.m.f. (c.f. e.m.f.) and into a circuit, which has a magnetic permeance (c.f. admittance); the inverse of permeance is *reluctance*, R_g (c.f. resistance). The load circuit may comprise a number of components in parallel, such as a leakage gap and a main gap, and because these experience the same m.m.f., Equation (4.50) shows that the permeance P_g used as the slope of the load line will be the sum of the component permeances:

$$P_g = P_1 + P_2 + P_3 + \cdots \tag{4.61}$$

The load circuit may also comprise a number of components in series, such as a pole piece and the main gap, and because these contain the same flux, the net load line slope will be the sum of the component reluctances:

$$\begin{aligned} P_g^{-1} &= P_1^{-1} + P_2^{-1} + P_3^{-1} + \cdots \\ R_g &= R_1 + R_2 + R_3 + \cdots \end{aligned} \tag{4.62}$$

As a simple example of a solution for the circuit flux levels, consider the magnetic circuit of Figure 4.20, which is similar to that of Figure 4.4 except for the addition of a movable armature. The soft iron pole pieces and armature are again assumed to be infinitely permeable ($H = 0$), and all the magnet flux is directed into the *main* gap (P_1) or the *leakage* gap (P_2), which are in parallel. It is further assumed that flux passes directly across the gaps as shown, so their permeanances can be easily calculated. Flux actually crosses the main gap twice, so its area is that between *one* pole piece and the armature, while its length is *twice* the clearance. The armature is constrained to move perpendicular to the pole pieces with an equal clearance on each side. One limit of operation is with the armature in contact with the poles, when all the magnet flux passes across the closed main gap. The other limit is with the armature completely removed, so

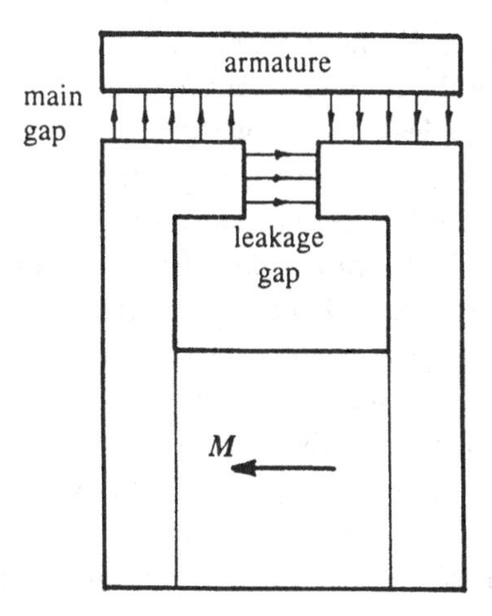

Figure 4.20. Magnetic circuit with a permanent magnet and parallel air gaps.

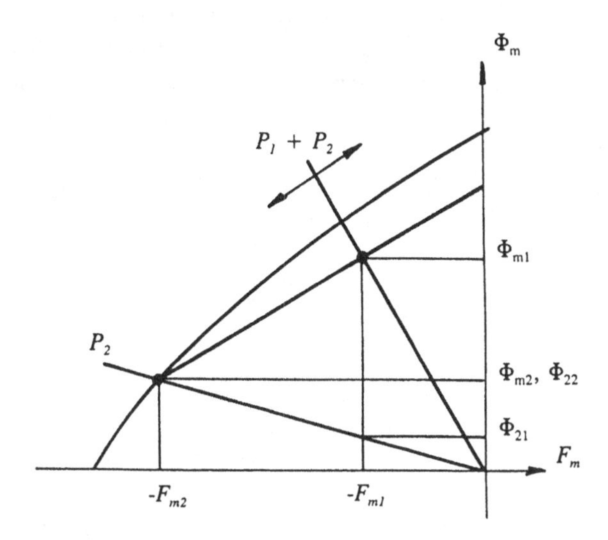

Figure 4.21. Operation of a magnet on a recoil line with parallel air gaps.

all the flux passes through the leakage gap. Movement of the armature between these limits causes the slope of the net load line to change, and drives the magnet along a recoil line as shown in Figure 4.21.

Because the gaps are in parallel, the load line slope given by Equation (4.61) is $P_1 + P_2$, becoming just P_2 when the armature is away and $P_1 \rightarrow 0$. The leakage gap effectively acts as a "reservoir" for the magnetic flux, preventing the magnet experiencing too great a demagnetization on removal of the armature. Without the armature, the magnet operates with a single air gap as described earlier at flux Φ_{m2} and m.m.f. $-F_{m2}$, delivering flux Φ_{22} $(=\Phi_{m2})$ to the leakage gap with an energy density of $\frac{1}{2}B_{m2}H_{m2}$. With the armature in place, the magnet operates at Φ_{m1}, $-F_{m1}$ as shown in Figure 4.21 and delivers an energy density of $\frac{1}{2}B_{m1}H_{m1}$ to the combined gaps. Being in parallel, the gaps share the same m.m.f., namely $-F_{m1}$ applied by the magnet, the leakage gap receiving a flux of Φ_{21} and the main gap a flux of $\Phi_{11}=\Phi_{m1}-\Phi_{21}$. This graphical solution therefore completely defines the flux distribution in the circuit for any given air gap dimensions. The proportion of the magnet's energy delivered to the *main* gap is maximized on the recoil line when $-F_m=-\frac{1}{2}F_{m2}$.

With the assumption that flux passes directly across the space between two co-planar surfaces, the length and area are easily defined for the calculation of permeance via Equation (4.60). There are some other common assumptions that are made to allow permeance to be determined by a simple formula. One is the passage of flux between two pole pieces, which subtend an angle α, the flux following arcs as shown in Figure 4.22. This is most likely to be an air gap, for which $\mu=1$, and commonly used values for α are 90° ($\frac{1}{2}\pi$) and 180° (π). Consider an arc strip of the gap at radius r, of width w and thickness δr, which has a path length $l_g=\alpha r$ and elemental area $\Delta A_g=w\,\delta r$. P_g is found by integrating the elemental permeance between the radial boundaries r_1 and r_2:

$$P_g=\int_{r_1}^{r_2}\frac{\mu_0 w}{\alpha r}\,\mathrm{d}r$$

$$=\frac{\mu_0 w}{\alpha}\ln\left(\frac{r_2}{r_1}\right) \tag{4.63}$$

An important variation on this type of path is the flux that leaks across the thickness l_m of a magnet *itself*. Consider the paths along one side of width w of the magnet shown in Figure 4.23, for which $\alpha=\pi$, $r_1=0$ and $r_2=\frac{1}{2}l_m$. While the full magnet m.m.f. acts upon paths originating beyond its faces, this is not true for leakage within its thickness – the arc strip at radius r experiences a magnet m.m.f. of $F_g=H_m\cdot 2r$ $(<H_m l_m)$. According to Equation (4.50), the flux in this strip is $\Delta\Phi_g=F_g\,\Delta P_g$, from which

$$\Phi_g=\int_0^{l_m/2}H_m\cdot 2r\,\frac{\mu_0 w}{\pi r}\,\mathrm{d}r$$

$$=\frac{\mu_0 w}{\pi}H_m l_m \tag{4.64}$$

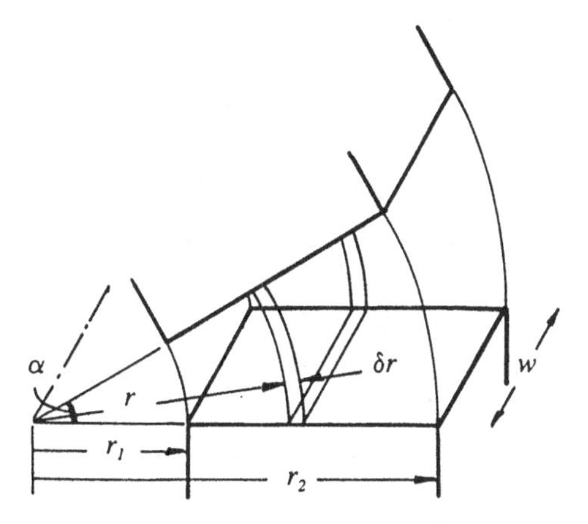

Figure 4.22. Flux between pole pieces at angle α.

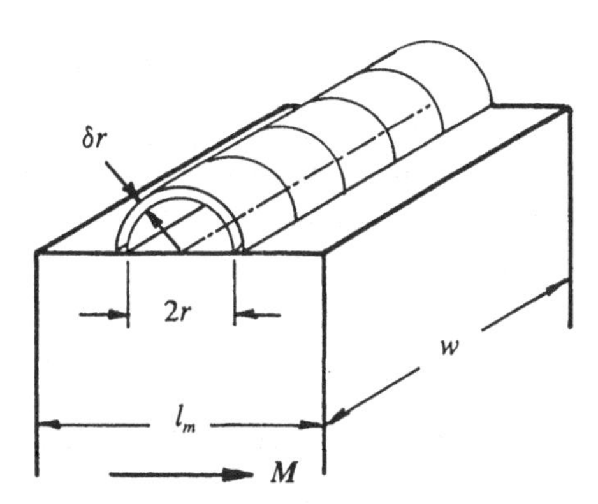

Figure 4.23. Flux outside a magnet edge.

When modeling the leakage flux across the *edge* of a magnet in an equivalent circuit, one may therefore locate a permeance across the *faces* of the magnet, with a value of

$$P_g = \frac{\mu_0 w}{\pi} \tag{4.65}$$

In another common configuration, flux passes radially between two pole pieces, which subtend an angle α as shown in Figure 4.24. This represents an air gap when $\mu = 1$, and two concentric cylinders when α is 360° (2π). A general calculation for the permeance of this radial air gap

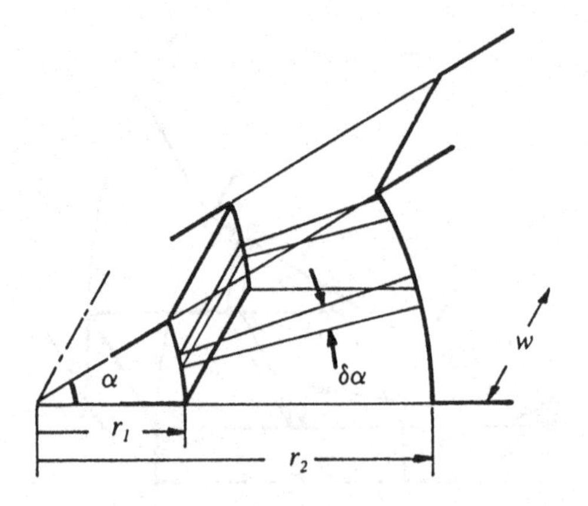

Figure 4.24. Flux between two concentric pole pieces.

is found to be

$$P_g = \frac{\mu_0 w \alpha}{\ln (r_2/r_1)} \tag{4.66}$$

An equivalent magnetic circuit of a device provides a straightforward model, which may be used to optimize its performance, but this is only achieved by simplifying the flux paths. Straight lines and arcs, or combinations thereof, are used as best estimates of the actual flux paths, and with some experience in their choice, the designer will achieve a simulation of the device, which is acceptable for sensitivity analysis and for preliminary selection of component dimensions and material properties.

References

Shercliff, J. A. (1977). *Vector Fields*. Cambridge: Cambridge University Press.
Woodson, H. H. and Melcher, J. R. (1968). *Electromechanical Dynamics, Part I: Discrete Systems*. New York: John Wiley & Sons.

5
Magnetic field analysis

5.1 Introduction

Using simplifications for the actual paths followed by flux in a magnetic device provides an approximate model as described in Chapter 4, which is useful both for preliminary selection of component materials and dimensions and for performing sensitivity analyses. Before building a prototype device, however, it is often desirable to perform a more detailed analysis of the flux distribution, to investigate the validity of the prior assumptions, and perhaps also to account for effects such as saturation and eddy currents. Depending upon the complexity of the design and the nature of the effects to be studied, a more accurate analytical solution of the field distribution may be attempted directly, or with the aid of commercially available computer software. The objective in this chapter is to provide the basis for the representation of magnetic fields in complex geometries, and an understanding of the most common techniques that are presently available.

The nature of the fields that occur in electromechanical devices may be categorized into three levels of complexity. The most straightforward are *magnetostatic* fields, which result from excitation at zero frequency. Nevertheless, materials may still be represented by non-linear characteristics in magnetostatic field solutions, as in the case of saturation. The next level of complexity involves alternating current excitation, which adds the effects of *eddy currents* to the field solution. These may be induced in conducting components and will diffuse into them, the result being modification of the field distribution and additional contributions to the power losses and forces in the device. Such a *skin-effect* may occur in the current-carrying excitation conductors themselves, the current being forced further towards the "skin" of the conductor as the frequency is raised, thus increasing the impedance of the component. Eddy currents are not only caused by fixed frequencies, but by transient changes in the excitation, as when a permanent magnet is charged from an "impulse" magnetizer. The third level of complexity, which is rarely encountered in electromechanical devices, involves coupling between electric and magnetic fields at high frequencies to radiate power into free space – this is more commonly used in microwave devices, antennae, radio transmitters, etc.

5.2 Magnetostatic fields

An analytical model of the field distribution in a device requires its representation by various of *Maxwell's Equations*, which were introduced as such in Section 4.2. *Ampère's Law* was expressed in integral form as Equation (4.3) and in differential form as

$$\nabla \times \boldsymbol{H} = \boldsymbol{J} \tag{1.27}$$

Maxwell recognized that, at high frequencies, it is necessary to modify the current density to include a "displacement" current as $(\boldsymbol{J} + \partial \boldsymbol{D}/\partial t)$, to account for the motion of electrical charges in space (Ramo, Whinnery and Van Duzer, 1965). The $\boldsymbol{D}$ term is insignificant at low frequencies and disappears when we consider only static fields. *Faraday's Law* was used in Equation (4.39) to relate the e.m.f. $-e$ induced in a coil to the rate of change of its flux linkage – around a single turn conductor linking flux Φ,

$$e = -\frac{\partial \Phi}{\partial t} \tag{5.1}$$

If the voltage per unit length of the conductor is defined as the *electric field E*, then Equation (5.1) may be rewritten for an elemental area $\mathrm{d}\boldsymbol{A}$, which is bounded by the loop $\mathrm{d}\boldsymbol{l}$, as

$$\oint \boldsymbol{E} \cdot \mathrm{d}\boldsymbol{l} = -\frac{\partial}{\partial t} \int \boldsymbol{B} \cdot \mathrm{d}\boldsymbol{A} \tag{5.2}$$

Applying *Stokes' Theorem* to $\boldsymbol{E}$ (as in Equation (4.2)) allows the area integration to be removed from Equation (5.2) and Faraday's Law to be written in differential form as

$$\nabla \times \boldsymbol{E} = -\frac{\partial \boldsymbol{B}}{\partial t} \tag{5.3}$$

Incidentally, this definition of $\boldsymbol{E}$ allows *Ohm's Law* for a conductor of resistivity ρ to be expressed as

$$\boldsymbol{E}=\rho\boldsymbol{J} \tag{5.4}$$

A model for a magnetic device will also always require invoking the principle of flux conservation, developed earlier as

$$\nabla\cdot\boldsymbol{B}=0 \tag{4.7}$$

We have represented a permanent magnet's characteristic in various ways thus far, each originating from Equation (1.26). For calculation of distributed fields, it is useful to describe the material's magnetization $\boldsymbol{M}$ in terms of its remanent value $\boldsymbol{M}_{\mathrm{r}}$ (when $\boldsymbol{H}=0$) and a *susceptibility* χ, which will itself be a function of $\boldsymbol{H}$ if the intrinsic characteristic is non-linear:

$$\boldsymbol{M}=\chi\boldsymbol{H}+\boldsymbol{M}_{\mathrm{r}} \tag{5.5}$$

In Equation (1.26),

$$\boldsymbol{B}=\mu_0[(1+\chi)\boldsymbol{H}+\boldsymbol{M}_{\mathrm{r}}] \tag{5.6}$$

Clearly, the relative permeability μ ($\geqslant 1$) is directly related to the susceptibility χ ($\geqslant 0$) via

$$\mu=1+\chi \tag{5.7}$$

In a *soft* magnetic material, which retains no magnetization ($\boldsymbol{M}_{\mathrm{r}}=0$), Equation (5.6) reduces to its familiar form:

$$\begin{aligned}\boldsymbol{B}&=\mu_0(1+\chi)\boldsymbol{H}\\&=\mu_0\mu\boldsymbol{H}\end{aligned} \tag{5.8}$$

We now have a fundamental set of equations, which form an analytical model of a magnetic device for the purpose of determining its field distribution – Equations (1.27), (4.7), (5.3), (5.4) and (5.6). The most common simplification is to solve for *magnetostatic* fields, ignoring any time-varying component of excitation that causes eddy currents; Equation (5.3) then becomes $\nabla\times\boldsymbol{E}=0$, although it is unnecessary even to calculate an electric field distribution. The magnetostatic field is completely defined by the remaining equations, though it is difficult to handle them in their present first-order differential forms; $\boldsymbol{H}$ and $\boldsymbol{M}$ can also be discontinuous. We therefore define either a scalar or a vector potential function that is both continuous and allows the field to be described by second-order differential equations, which are more easily solved.

We may express flux density in terms of a *scalar potential* via

$$\boldsymbol{B} = -\mu_0 \nabla \psi \tag{5.9}$$

The *grad* operation that is performed on this scalar quantity clearly produces the vector $\nabla\psi$, which is written using unit vectors $\boldsymbol{i}, \boldsymbol{j}$ and $\boldsymbol{k}$ in the x, y and z directions respectively as

$$\nabla \psi = \boldsymbol{i}\frac{\partial \psi}{\partial x} + \boldsymbol{j}\frac{\partial \psi}{\partial y} + \boldsymbol{k}\frac{\partial \psi}{\partial z} \tag{5.10}$$

Consequently, the three Cartesian components of $\boldsymbol{B}$ are

$$B_x = -\mu_0 \frac{\partial \psi}{\partial x}; \qquad B_y = -\mu_0 \frac{\partial \psi}{\partial y}; \qquad B_z = -\mu_0 \frac{\partial \psi}{\partial z} \tag{5.11}$$

Combining Equations (4.7) and (5.9) produces the desired second-order differential equation in scalar magnetic potential. Since μ_0 is a constant,

$$\nabla \cdot \nabla \psi = 0 \tag{5.12}$$

This *div grad* operation on ψ is abbreviated to ∇^2, and what is known as *Laplace's Equation* is written as

$$\nabla^2 \psi = 0 \tag{5.13}$$

The complete form of the ∇^2 operation may be expressed by substituting the components of $\boldsymbol{B}$ from Equation (5.11) into Equation (4.6), giving

$$\frac{\partial^2 \psi}{\partial x^2} + \frac{\partial^2 \psi}{\partial y^2} + \frac{\partial^2 \psi}{\partial z^2} = 0 \tag{5.14}$$

Once this second-order differential equation has been solved, the $\boldsymbol{B}$-distribution can be determined using Equation (5.9). However, this solution will also require boundary conditions, which include the excitation sources. For permanent magnets, this is the intrinsic magnetization $\boldsymbol{M}$, which can be incorporated into the scalar potential solution by combining Equations (1.26) and (5.9) as

$$\boldsymbol{H} = -\nabla \psi - \boldsymbol{M} \tag{5.15}$$

It is certainly desirable to represent the *magnetizing force* by a scalar potential also, though this may only be ψ in a medium where $\boldsymbol{M}$ is absent. A new scalar potential is therefore defined via

$$\nabla \Psi = \nabla \psi + \boldsymbol{M} \tag{5.16}$$

Even within a permanent magnet, $\boldsymbol{H}$ may now be derived from Ψ using

$$\boldsymbol{H} = -\nabla\Psi \tag{5.17}$$

Applying the *div grad* operation to both sides of this, and noting Equation (5.12), we derive a more general second-order differential equation in scalar potential, known as *Poisson's Equation*:

$$\nabla^2\Psi = \nabla\cdot\boldsymbol{M} \tag{5.18}$$

This reduces to the simpler form of *Laplace's Equation* (5.13) whenever $\nabla\cdot\boldsymbol{M}=0$. This condition clearly exists in all regions external to a magnet where $\boldsymbol{M}$ does not exist, but it also applies within a magnet's volume if the magnetization is uniform. This was illustrated in Figure 1.4 by an equivalent current density $\boldsymbol{J}_{\mathrm{m}}$ only on the magnet's sides, but $\nabla\cdot\boldsymbol{M}\neq 0$ occurs only on the end faces of the magnet representing the *North* and *South* poles. It is common to assume that a permanent magnet is uniformly magnetized, because otherwise the solution of Equation (5.18) is more complex.

We have shown that magnet surfaces contain discontinuities of $\boldsymbol{M}$. In solving the field equations, however, we must also observe some general boundary conditions at interfaces between regions with dissimilar magnetic properties. Conservation of flux across an interface must be observed, as it was for the volume element in Figure 4.2 and Equation (4.4). Imagine that this element shrinks to enclose a surface between two regions as shown in Figure 5.1 ($\delta y \to 0$). The side areas of the element $\delta x\cdot\delta y$ and $\delta y\cdot\delta z$ are now very small relative to the surface area $\delta x\cdot\delta z$, such that flux only enters and leaves the element through these two faces. Hence, across an interface, the *normal* component of flux density B_{n} is continuous:

$$B_{\mathrm{n}1} = B_{\mathrm{n}2} \tag{5.19}$$

Similarly, Equation (4.3) may be applied along a closed path, which straddles the surface as shown in Figure 5.1, but as $\delta y \to 0$, the only contributions to $\oint \boldsymbol{H}\cdot\mathrm{d}\boldsymbol{l}$ come along lines of length δl parallel to the surface plane. Hence, across an interface, the *tangential* component of magnetizing force H_{t} is governed by

$$H_{\mathrm{t}1} - H_{\mathrm{t}2} = \frac{i}{\delta l} \tag{5.20}$$

H_{t} is discontinuous by virtue of the current per unit width flowing in the surface plane – this includes *equivalent* currents. Only when surface currents are not present will H_{t} be continuous across the interface.

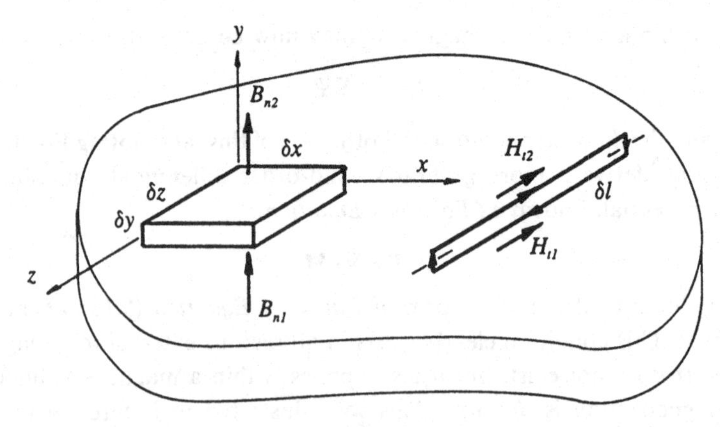

Figure 5.1. Magnetic fields at an interface between two different magnetic media.

Scalar potential provides a straightforward method of representation for magnetizing force or flux density. When it comes to solving the field equations, it is an attractive feature that this potential has a single numerical value at each location throughout a magnetic system. However, the discontinuity caused by surface boundary currents adds significant complexity to the solution, and it is equally troublesome to represent excitation current sources. Consider the most obvious way to introduce $\boldsymbol{J}$ as a scalar potential, by combining Equations (1.27) and (5.17):

$$\nabla \times \nabla \Psi = -\boldsymbol{J} \tag{5.21}$$

Unfortunately, Ψ is lost from this equation because the $\nabla \times \nabla$ operation applied to a scalar quantity always equals zero. The components of $\nabla \times \boldsymbol{M}$ were written in Equation (1.18), and those of $\nabla \psi$ were given in Equation (5.10); by comparison, $\nabla \times \nabla \Psi$ is

$$\boldsymbol{i}\left(\frac{\partial \Psi}{\partial y\,\partial z}-\frac{\partial \Psi}{\partial z\,\partial y}\right)+\boldsymbol{j}\left(\frac{\partial \Psi}{\partial z\,\partial x}-\frac{\partial \Psi}{\partial x\,\partial z}\right)+\boldsymbol{k}\left(\frac{\partial \Psi}{\partial x\,\partial y}-\frac{\partial \Psi}{\partial y\,\partial x}\right)=0 \tag{5.22}$$

Because of these difficulties, it is sometimes advantageous to solve a magnetostatic field problem using the *vector potential* $\boldsymbol{A}$, from which flux density is defined by

$$\boldsymbol{B} = \nabla \times \boldsymbol{A} \tag{5.23}$$

Because Equation (4.7) indicates that $\nabla \cdot (\nabla \times \boldsymbol{A}) = 0$, a further condition is required to define $\boldsymbol{A}$ completely, which is chosen to be

$$\nabla \cdot \boldsymbol{A} = 0 \tag{5.24}$$

The contributions of both $\boldsymbol{M}$ and $\boldsymbol{J}$ are included in the calculation of $\boldsymbol{B}$ by taking the *curl* of each term in Equation (5.6), and substituting Equation (1.27):

$$\nabla \times \frac{1}{\mu_0 \mu} \boldsymbol{B} = \boldsymbol{J} + \nabla \times \frac{\boldsymbol{M}_r}{\mu} \tag{5.25}$$

Unlike the operation on a scalar quantity, these terms are retained in the second-order differential equation in vector potential derived by combining Equations (5.23) and (5.25):

$$\nabla \times \frac{1}{\mu_0 \mu} \nabla \times \boldsymbol{A} = \boldsymbol{J} + \nabla \times \frac{\boldsymbol{M}_r}{\mu} \tag{5.26}$$

There are various alternative representations of the magnetic properties, which may be more amenable to solving the field equations. In a permanent magnet, its relative permeability μ and remanent magnetization $\boldsymbol{M}_r$ may be treated as an equivalent current density $\boldsymbol{J}_m$ and incorporated within $\boldsymbol{J}$ as

$$\boldsymbol{J}_m = \nabla \times \frac{\boldsymbol{M}_r}{\mu} \tag{5.27}$$

This simplifies Equation (5.26) to the form of Equation (5.28). In a soft magnetic material for which $\boldsymbol{M}_r$ and $\boldsymbol{J}_m$ are zero, it is always true that

$$\nabla \times \frac{1}{\mu_0 \mu} \nabla \times \boldsymbol{A} = \boldsymbol{J} \tag{5.28}$$

In an air gap region, a further simplification to Equation (5.28) will be $\mu = 1$, so

$$\frac{1}{\mu_0} \nabla \times \nabla \times \boldsymbol{A} = \boldsymbol{J} \tag{5.29}$$

By writing out the components of the vector operations in full, as we have illustrated before, it may be proven that the following identity exists for any vector, in this case $\boldsymbol{A}$:

$$\nabla \times \nabla \times \boldsymbol{A} = \nabla \nabla \cdot \boldsymbol{A} - \nabla^2 \boldsymbol{A} \tag{5.30}$$

Combining Equations (5.24), (5.29) and (5.30) yields *Poisson's Equation* in vector potential:

$$\nabla^2 \boldsymbol{A} = -\mu_0 \boldsymbol{J} \tag{5.31}$$

Formulating a second-order differential equation for magnetostatic fields using vector potential clearly facilitates inclusion of excitation current sources. The boundary conditions for $\boldsymbol{B}$ and $\boldsymbol{H}$, which occur at interfaces between dissimilar media, are still as described by Equations (5.19) and (5.20). The only major drawback occurs in the solution of three-dimensional field problems, where it is a much more cumbersome task to manipulate three sets of vector components, rather than a single scalar potential function. Visualization of the field distribution by the flux lines is easier with $\boldsymbol{A}$ than with ψ or Ψ. The definition of $\boldsymbol{A}$ by Equation (5.24) means that flux lines are *coincident* with equipotentials, whereas flux lines are *perpendicular* to scalar equipotentials.

5.3 The finite difference method

Having developed the equations to represent magnetic fields, a basic review of the most common techniques that are used to solve them will also be provided. To make this understandable, only uncomplicated cases will be considered, as with our limitation to magnetostatic conditions. Except in the case of simple geometries and boundary conditions, it is almost impossible to achieve direct analytical solution of the second-order differential equations. In practical electromechanical devices, we frequently encounter complex shapes and non-linear material properties, for which numerical methods usually offer the best solution.

The *finite difference* method is the most elementary technique for a designer to implement. Its underlying principle involves replacing all derivatives in the differential equations with finite difference expressions, which provide good approximations to the derivatives. As an example we shall use Equation (5.28), but for clarity we will only develop a simplified solution for $\boldsymbol{A}$, with $\boldsymbol{B}$ confined to an x–y plane and caused by a uniform, uni-directional excitation current $\boldsymbol{J}=J_z$. In this two-dimensional problem, there are no variations in the parameters in the z direction. Following the form of Equation (1.18), the components of $\boldsymbol{B}$ are

$$\boldsymbol{B}=\nabla\times\boldsymbol{A}=\boldsymbol{i}\left(\frac{\partial A_z}{\partial y}-\frac{\partial A_y}{\partial z}\right)+\boldsymbol{j}\left(\frac{\partial A_x}{\partial z}-\frac{\partial A_z}{\partial x}\right)+\boldsymbol{k}\left(\frac{\partial A_y}{\partial x}-\frac{\partial A_x}{\partial y}\right) \quad (5.32)$$

In our example, there is neither a $\boldsymbol{k}$ component nor any $\partial/\partial z$ variations, so

$$\frac{1}{\mu}\nabla\times\boldsymbol{A}=\boldsymbol{i}\frac{1}{\mu}\frac{\partial A_z}{\partial y}-\boldsymbol{j}\frac{1}{\mu}\frac{\partial A_z}{\partial x} \quad (5.33)$$

Notice that, under these conditions, $\boldsymbol{B}$ is only represented by the A_z component, which henceforth we shall write as $\boldsymbol{A}$ for economy. Now, applying the *curl* operation again yields

$$\nabla \times \frac{1}{\mu} \nabla \times \boldsymbol{A} = -\boldsymbol{i} \frac{\partial}{\partial z}\left(-\frac{1}{\mu}\frac{\partial A}{\partial x}\right) + \boldsymbol{j}\frac{\partial}{\partial z}\left(\frac{1}{\mu}\frac{\partial A}{\partial y}\right)$$

$$+\boldsymbol{k}\frac{\partial}{\partial x}\left(-\frac{1}{\mu}\frac{\partial A}{\partial x}\right) - \boldsymbol{k}\frac{\partial}{\partial y}\left(\frac{1}{\mu}\frac{\partial A}{\partial y}\right) \tag{5.34}$$

Because there is no $\partial/\partial z$ variation, only the $\boldsymbol{k}$-directed z component of Equation (5.28) remains, in accord with the J_z excitation:

$$\frac{\partial}{\partial x}\left(\frac{1}{\mu}\frac{\partial A}{\partial x}\right) + \frac{\partial}{\partial y}\left(\frac{1}{\mu}\frac{\partial A}{\partial y}\right) = -\mu_0 J_z \tag{5.35}$$

For numerical solution, the partial derivatives are to be replaced by difference approximations. The x–y plane is divided into the grid shown in Figure 5.2, with spacings δx and δy, and for illustration consider just the five elements that are marked. Each grid element has a unique value of $\boldsymbol{A}$, and between adjacent elements there is a single permeability μ. Provided that the spacings are sufficiently small, a first derivative may be approximated in the following manner:

$$\frac{\partial A}{\partial x} = \frac{\boldsymbol{A}_{(x+\delta x/2)} - \boldsymbol{A}_{(x-\delta x/2)}}{\delta x} \tag{5.36}$$

Likewise, referring to the convention in Figure 5.2, the second derivatives about the central element containing $\boldsymbol{A}_0$ may be written in the form

$$\frac{\partial}{\partial x}\left(\frac{1}{\mu}\frac{\partial A}{\partial x}\right) = \frac{\frac{1}{\mu_1}\left(\frac{\boldsymbol{A}_1 - \boldsymbol{A}_0}{\delta x}\right) - \frac{1}{\mu_3}\left(\frac{\boldsymbol{A}_0 - \boldsymbol{A}_3}{\delta x}\right)}{\delta x}$$

$$\frac{\partial}{\partial y}\left(\frac{1}{\mu}\frac{\partial A}{\partial y}\right) = \frac{\frac{1}{\mu_2}\left(\frac{\boldsymbol{A}_2 - \boldsymbol{A}_0}{\delta y}\right) - \frac{1}{\mu_4}\left(\frac{\boldsymbol{A}_0 - \boldsymbol{A}_4}{\delta y}\right)}{\delta y} \tag{5.37}$$

These constitute the left-hand side of Equation (5.35), the excitation at the central element being

$$\boldsymbol{J}_0 = \boldsymbol{J}_{z(x+\delta x/2, y+\delta y/2)} \tag{5.38}$$

Let the spacings δx and δy be identical ($=\delta$), and substitute Equations (5.37) and (5.38) into (5.35). The terms may then be rearranged so that $\boldsymbol{A}_0$ is calculated from the neighboring potentials, the adjoining

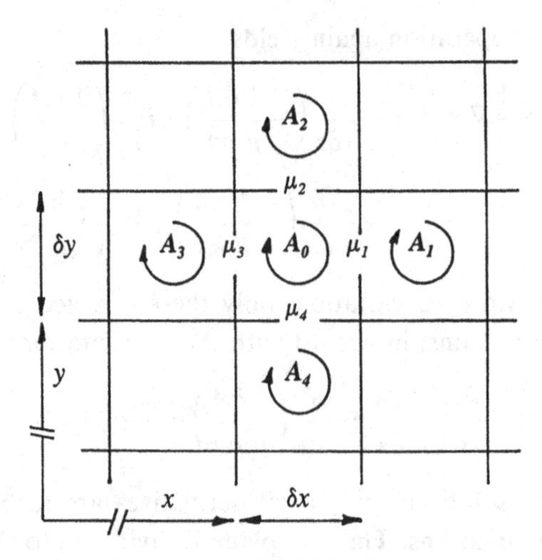

Figure 5.2. Elemental grid in the x–y plane.

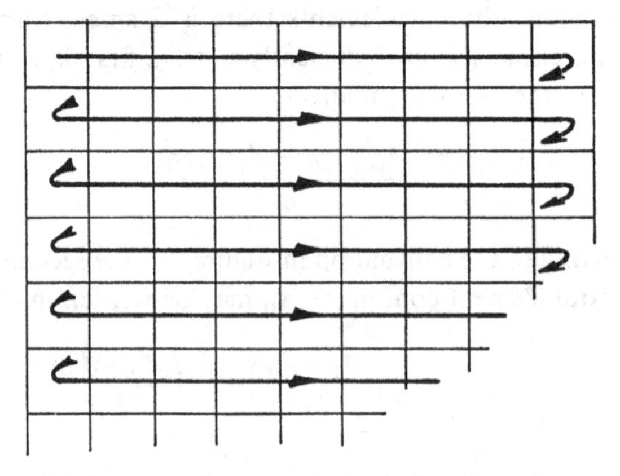

Figure 5.3. Sequence for numerical solution throughout a grid.

permeabilities and its own excitation source as

$$A_0 = \frac{\left(\sum_{i=1}^{4} \frac{A_i}{\mu_i}\right) + \delta^2 \mu_0 J_0}{\sum_{i=1}^{4} \frac{1}{\mu_i}} \tag{5.39}$$

The numerical solution is performed sequentially throughout the entire grid, a possible sequence for the scan being shown in Figure 5.3. Provided that any reasonable initial values for A are chosen, successive scans will produce progressively better estimates of the true potentials. Comparing Figures 5.2 and 5.3, it is clear that on the $(n+1)$th iteration, the new value for A_0 uses the current $(n+1)$th values from elements 1 and 4, but the previous nth values from elements 2 and 3. On the $(n+1)$th iteration, Equation (5.39) is used to recalculate A_0^{n+1} from its neighbors according to

$$A_0^{n+1} = \frac{\left(\frac{A_1}{\mu_1}\right)^n + \left(\frac{A_2}{\mu_2}\right)^{n+1} + \left(\frac{A_3}{\mu_3}\right)^{n+1} + \left(\frac{A_4}{\mu_4}\right)^n + \delta^2 \mu_0 J_0}{\left(\frac{1}{\mu_1}\right)^n + \left(\frac{1}{\mu_2}\right)^{n+1} + \left(\frac{1}{\mu_3}\right)^{n+1} + \left(\frac{1}{\mu_4}\right)^n} \tag{5.40}$$

This numerical technique obviously provides a straightforward method of solving the second-order differential equation, and is easily implemented using a computer. If the solution to the field problem is converging with successive iterations (Smith, 1965), then the difference between the $(n+1)$th and nth values of A will eventually become negligible for *all* elements in the grid. This difference is known as the *residual*, expressed for each element as

$$R_0^{n+1} = A_0^{n+1} - A_0^n \tag{5.41}$$

When the maximum residual has become sufficiently small, i.e. all residuals $R^{n+1} \to 0$, then all potentials A^{n+1} have reached their true values and no further iterations are necessary – a solution to Equation (5.35) has been achieved. As a practical matter, it is more usual to calculate R_0^{n+1} through the grid by combining Equations (5.40) and (5.41), rather than A_0^{n+1}, which is then calculated from

$$A_0^{n+1} = A_0^n + R_0^{n+1} \tag{5.42}$$

Because the residual represents the *error* in the calculation at each element, it is possible to accelerate the numerical procedure and reduce the number of iterations by using a procedure known as *over-relaxation* (Smith, 1965); the residual is multiplied by a *relaxation factor* ω (a constant in the range $1 < \omega < 2$) and Equation (5.42) is replaced by

$$A_0^{n+1} = A_0^n + \omega R_0^{n+1} \tag{5.43}$$

Interfaces between dissimilar media are accounted for when using Equation (5.40), though a grid line should be chosen to coincide with the boundary. However, values of A adjacent to a boundary may also be used to satisfy the continuity conditions for normal $\boldsymbol{B}$ and tangential $\boldsymbol{H}$ derived as Equations (5.19) and (5.20).

5.4 Permanent magnet model

A permanent magnet is modeled after Equation (5.6) by the permeability function μ throughout its volume and the remanent magnetization $\boldsymbol{M}_r$, which appears only on its surface if the magnetization is uniform,

$$\boldsymbol{B} = \mu_0(\mu \boldsymbol{H} + \boldsymbol{M}_r) \tag{5.44}$$

For the simplified solution of Equation (5.28) developed in the previous section, the uniform magnetization in a typical anisotropic material may be taken to lie in the x direction as shown in Figure 5.4. When represented as an equivalent current density via Equation (5.27), the remaining terms for the excitation are

$$\boldsymbol{J}_m = \boldsymbol{j}\frac{\partial\left(\dfrac{\boldsymbol{M}_r}{\mu}\right)}{\partial z} - \boldsymbol{k}\frac{\partial\left(\dfrac{\boldsymbol{M}_r}{\mu}\right)}{\partial x} \tag{5.45}$$

This is simplified further in the two-dimensional problem, for which $\boldsymbol{J}_m$ has its J_z component only – the equivalent currents over the upper and lower sides of the magnet are removed in Figure 5.4:

$$J_z = -\boldsymbol{k}\frac{\partial\left(\dfrac{\boldsymbol{M}_r}{\mu}\right)}{\partial x} \tag{5.46}$$

In the finite difference formulation for vector potential, the source terms in Equations (5.39) and (5.40) may be rewritten in the case of a magnet as

$$\delta^2 \mu_0 J_0 = \delta \mu_0 \frac{\boldsymbol{M}_r}{\mu} \tag{5.47}$$

As depicted in Figure 1.4, the magnetization is represented as current in side boundary regions of width δ. The magnet is similar to a conventional

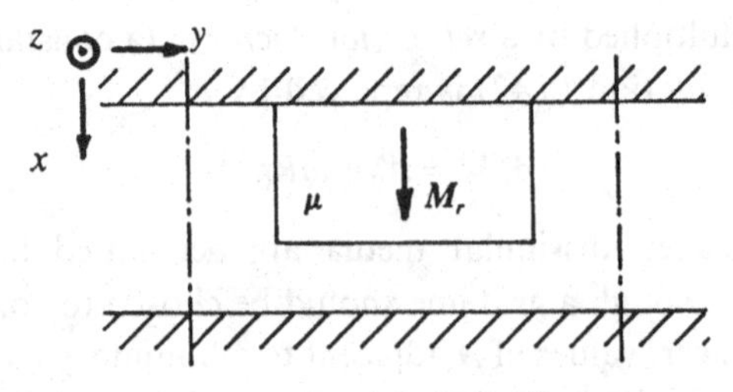

Figure 5.4. Two-dimensional approximation of a uniformly magnetized permanent magnet.

solenoid, in which real conductor current is divided amongst the grid elements it spans according to $i_0 = \delta^2 J_0$.

The demagnetization characteristic of a permanent magnet provides values for μ, χ and M_r along a given axis, usually the preferred direction in an anisotropic material. It has already been noted that μ and χ will be functions of H if the characteristic is non-linear. As has been deduced from Equation (4.25), over a range where M is constant, $\chi = 0$, $\mu = 1$, $\mu_{rec} = 1$ and $M_r = M_{sat}$. For a magnet operating under conditions where M is not constant, the demagnetization characteristic will be required to provide updated values for μ and M_r at each grid element. While Equation (5.40) is calculated for each element in one iteration through the grid, such non-linearity also requires determination of B and A and correction for μ and M_r within each iteration, a significantly more time-consuming procedure. Fortunately, it is a reasonable assumption for most ceramic ferrite and rare earth magnets operating under normal conditions that a linear characteristic may be used to provide unique values for μ and M_r. This does not force B and H to be aligned with M within a magnet, as shown in Figure 4.3 – it only requires that M be pinned in the given direction described by the characteristic.

In some materials, however, a localized self-demagnetizing field H at angle θ_0 to the preferred direction will cause M to rotate by an angle θ, as shown in Figure 1.15. Materials based upon shape anisotropy are particularly susceptible to this effect, and the torque rotating the magnetization can be determined by differentiating the energy function Equation (1.58). This technique is not well suited to incorporation in a numerical field solution, but it has been found possible to utilize orthogonal magnetization characteristics, at least for alnico magnets (Campbell and Al-Murshid, 1982). Using subscripts p and t to designate preferred and transverse axes respectively, the intrinsic characteristics are represented by a pair of equations rather than by Equation (5.5) alone:

$$M_p = \chi_p H_p + M_r$$
$$M_t = \chi_t H_t \tag{5.48}$$

As discussed for recoil permeability, measurement of these two orthogonal characteristics provides approximately *constant* values for χ_p and χ_t, and hence μ_p and μ_t for use in numerical formulae such as Equation (5.40).

A solution for the field distribution using scalar rather than vector potential follows *Poisson's Equation* (5.18), where a permanent magnet's sources occur where $\nabla \cdot M \neq 0$, primarily at its *North* and *South* pole faces.

As we have noted, however, non-uniform material properties complicate the solution of this equation, for which the components of $\boldsymbol{M}$ must be separated. Using Equation (5.5) in Equation (5.15) yields

$$\mu \boldsymbol{H} = -\nabla\psi - \boldsymbol{M}_r \tag{5.49}$$

Applying the *div grad* operation to both sides and substituting Equation (5.17), we get a second-order differential equation in Ψ which includes both μ and $\boldsymbol{M}_r$:

$$\nabla \cdot \mu \nabla \Psi = \nabla \cdot \boldsymbol{M}_r \tag{5.50}$$

It has been demonstrated that current sources are very difficult to represent using scalar potential, though without these it is quite straightforward to replace the derivatives in Equation (5.50) with finite differences and perform numerical solution for Ψ. If a magnet has the uniform magnetization shown in Figure 5.4, then the source term reduces to pole distributions in end boundary regions of width δ, written as

$$\delta^2(\nabla \cdot \boldsymbol{M}_r) = \delta \boldsymbol{M}_r \tag{5.51}$$

5.5 The finite element method

It has been shown that magnetic field distributions can be determined by solving second-order partial differential equations in either scalar or vector potential, although the latter representation facilitates inclusion of excitation current sources. Numerical methods are generally required to obtain this solution for all but the simplest geometries, but use of the finite difference method can become tedious when modeling the complex shapes found in many electromechanical devices. The *finite element* method provides a technique to alleviate this difficulty, and furthermore, it allows inclusion of eddy currents directly into the field solution.

The principle of the finite element method involves transforming the field equations, such as Equation (5.28) in vector potential, into *energy functionals*, which are expressions having the dimensions of energy but which are not the actual field energies determined in earlier Chapters. The functional Π has a numerical value at each point within a region, and is an *integral representation* throughout the entire field volume Ω of the variables that are functions of the geometry, the material properties, the potential solution and its derivatives:

$$\Pi = \int_\Omega \mathrm{fn}\left(x, y, z, A, \frac{\partial A}{\partial x}, \frac{\partial A}{\partial y}, \frac{\partial A}{\partial z}, \ldots\right) \mathrm{d}\Omega \tag{5.52}$$

The condition for a stable solution to the field distribution is that minimization of this energy functional yields the original partial differential equation of the field. A suitable functional will therefore be any expression whose minimum, obtained by differentiation with respect to potential and being set equal to zero, is the field equation. Under these conditions, the field distribution may be determined from the energy functional rather than from the partial differential equation, a technique that affords the advantages already mentioned.

Derivation of appropriate forms for the energy functional is far beyond the scope of this text, although the suitability of the following expressions in vector and scalar potentials can be proven (Chari and Silvester, 1980):

$$\Pi = \int_{\Omega} [\boldsymbol{A}^{\mathrm{T}}\boldsymbol{L}\boldsymbol{A} - 2\boldsymbol{A}^{\mathrm{T}}\boldsymbol{f}]\, \mathrm{d}\Omega \tag{5.53}$$

$$\Pi = \int_{\Omega} [\Psi \boldsymbol{L} \Psi - 2\Psi \boldsymbol{f}]\, \mathrm{d}\Omega \tag{5.54}$$

These can be transformed into their respective partial differential equations, which have the following forms:

$$\boldsymbol{L}\boldsymbol{A} = \boldsymbol{f} \tag{5.55}$$

$$\boldsymbol{L}\Psi = \boldsymbol{f} \tag{5.56}$$

The operator $\boldsymbol{L}$ and function $\boldsymbol{f}$ are chosen to model the excitation sources that are present. For example, magnetostatic fields with current sources $\boldsymbol{J}$ are governed by Equation (5.28), which, compared with Equation (5.55), yields

$$\begin{aligned} \boldsymbol{L} &= \nabla \times \frac{1}{\mu_0 \mu} \nabla \times \\ \boldsymbol{f} &= \boldsymbol{J} \end{aligned} \tag{5.57}$$

In this case the energy functional to be minimized will be

$$\Pi = \int_{\Omega} \boldsymbol{A}^{\mathrm{T}} \nabla \times \frac{1}{\mu_0 \mu} \nabla \times \boldsymbol{A}\, \mathrm{d}\Omega - 2 \int_{\Omega} \boldsymbol{A}^{\mathrm{T}} \boldsymbol{J}\, \mathrm{d}\Omega \tag{5.58}$$

A permanent magnet material can be incorporated within $\boldsymbol{J}$ in the functional using Equation (5.27).

While use of Equation (5.58) is common for all types of two-dimensional field problems, its solution in three dimensions becomes unnecessarily complicated. Not only must the three components of $\boldsymbol{A}$ be handled within the functional, but so too must the requirement that $\nabla \cdot \boldsymbol{A}=0$ (Equation (5.24)). Accurate determination of field surrounding a permanent magnet frequently does require a three-dimensional model, for which it is simpler to use scalar potential Ψ and its functional in Equation (5.54) (Campbell, Chari and D'Angelo, 1981). The magnetostatic field from a permanent magnet using Ψ is governed by Equation (5.50), which compared to Equation (5.56) yields

$$\begin{aligned} L &= \nabla \cdot \mu \nabla \\ f &= \nabla \cdot \boldsymbol{M}_{\mathrm{r}} \end{aligned} \tag{5.59}$$

The energy functional to be minimized is

$$\Pi = \int_{\Omega} \Psi \nabla \cdot \mu \nabla \Psi \mathrm{d}\Omega - 2 \int_{\Omega} \Psi \nabla \cdot \boldsymbol{M}_{\mathrm{r}} \, \mathrm{d}\Omega \tag{5.60}$$

If the remanent magnetization $\boldsymbol{M}_{\mathrm{r}}$ is uniform within the magnet volume such that $\nabla \cdot \boldsymbol{M}_{\mathrm{r}}=0$ everywhere except on the pole faces, then this energy functional can be considerably simplified to one integration through the entire region Ω plus another over the magnet's surface γ (Campbell, Chari and D'Angelo, 1981):

$$\Pi = -\int_{\Omega} \mu |\nabla \Psi|^2 \, \mathrm{d}\Omega - 2\boldsymbol{M}_{\mathrm{r}} \int_{\gamma} \Psi \, \mathrm{d}\gamma \tag{5.61}$$

The relative permeability μ still exists throughout the entire region, providing a model for all magnetic materials, soft or hard, linear or non-linear.

Having defined the field problem by a partial differential equation and obtained a suitable energy functional for this, *finite element* solution of the field distribution is initiated by dividing the entire region into elements. This need not be a rectangular grid, as in Figure 5.2 for the finite difference method, but may comprise triangular elements of various shapes as shown in Figure 5.5 – these element boundaries adapt much more readily to complex geometries (for three-dimensional solution, the elements are shaped as triangular prisms). Within each element, the potential is described according to its nodal values, as illustrated in Figure 5.5 and written generally as either of

$$\boldsymbol{A} = \sum_{i=a,b,c} \zeta_i \boldsymbol{A}_{\mathrm{i}} \tag{5.62}$$

$$\Psi = \sum_{i=a,b,c} \zeta_i \Psi_i \tag{5.63}$$

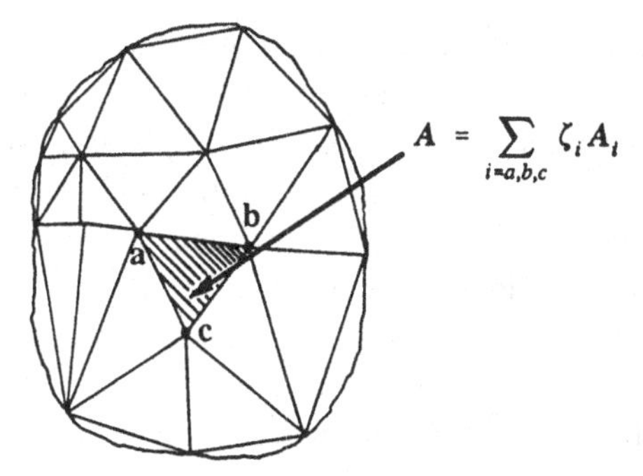

Figure 5.5. Subdivision of a region into triangular finite elements with their vector potential function.

Each nodal potential value is multiplied by a weighting function ζ_i, which accounts for dimensional and property changes throughout the mesh, much as did μ_i, δx and δy in Equation (5.37). The energy functional at each element is expressed in terms of its nodal potentials, by substituting Equation (5.62) into (5.58), or Equation (5.63) into (5.60) (or (5.61)). Each functional is then minimized by differentiating it with respect to its potential and setting it equal to zero, i.e.

$$\frac{\partial \Pi}{\partial A}=0 \tag{5.64}$$

$$\frac{\partial \Pi}{\partial \Psi}=0 \tag{5.65}$$

This procedure is performed for each nodal potential at each element, resulting in a set of linear algebraic equations, which describe the entire region. Solution of such an equation set is direct and relatively straightforward to implement, yielding a complete potential distribution throughout the region from which the flux distribution may be derived. The finite element method is iterative in nature because an initial potential distribution must be assumed, which is successively improved until all the residuals are sufficiently small.

Our description of the finite element method has been necessarily qualitative, though a more thorough mathematical derivation can be found in specialist texts dealing with application to magnetic field problems (Chari and Silvester, 1980). The permanent magnet designer usually finds it more convenient to utilize one of several commercially available software packages, because they provide all the programming steps outlined above in addition to user-friendly means to detail the geometry. Software is available that addresses various levels of complexity, the most straightforward being the solution to magnetostatic fields in two dimensions. Even so, the designer may need to implement a model of the material according to one of the methods described herein.

We conclude this section with an example solution of the two-dimensional magnetostatic field within a linear actuator using the vector potential energy functional of Equation (5.58), since this is most commonly used in commercial software. As shown in Figure 5.6, an air-cored coil can travel along the central limb of a soft magnetic core (mechanical support not shown). The excitation current in this coil interacts with the fields from two identical permanent magnets, producing the translational force on the coil. We always look for ways to reduce the scope of the problem, and hence the magnitude of the calculation. An axis of symmetry

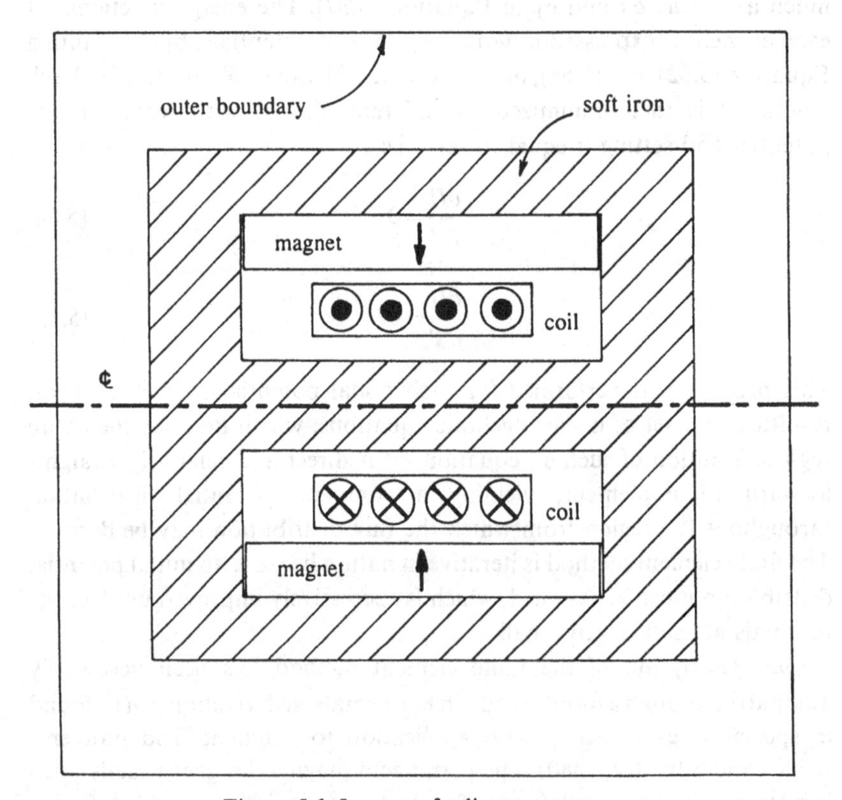

Figure 5.6. Layout of a linear actuator.

exists through the central limb as shown, across which no flux will pass; therefore, a field solution need only be performed for the upper (or lower) half of the complete region. Since our software provides a formulation of the energy functional, our first step is to define the geometry of the device, and allow the program to establish a finite element mesh. Figure 5.7 shows a relatively coarse mesh for clarity, with elements placed to coincide with component boundaries. Because finite elements are not required to have equal shapes or areas, it is possible to create a finer mesh in sectors where greater detail of the field distribution is needed, such as active air gaps. An outer boundary for the entire region is chosen at sufficient distance not to affect the flux distribution within the device, so it may therefore be an equi-potential with $A=0$. Likewise, no flux crosses the symmetry line, and according to Equation (5.23) it too must be an equi-potential – we may again choose $A=0$. We might also encounter symmetry conditions in which flux only crosses *normal* to the boundary (not shown anywhere in this example), and these may be established by allowing A to have equal values in neighboring elements across the boundary.

While the coil area carries a real excitation current density J, the permanent magnet is represented by an equivalent current density J_m, calculated from M_r and μ using Equation (5.27). Using the two-dimensional approximation shown in Figure 5.4, J_m is simplified to its J_z component in Equation (5.46), and with a magnet length of δx, this has

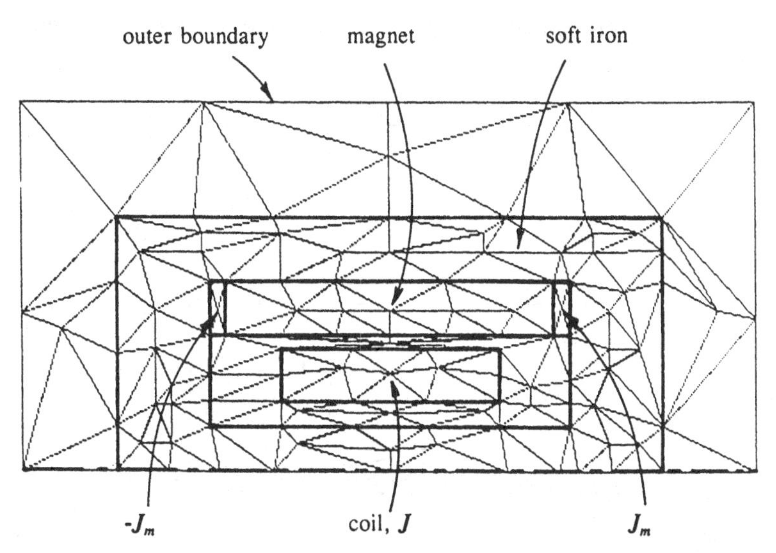

Figure 5.7. Coarse finite element mesh, with component outlines enhanced.

the value

$$J_z = \frac{M_r}{\mu \, \delta x} \tag{5.66}$$

While μ is distributed through the entire magnet, this current density exists only in its side boundary regions, which are shown in Figure 5.7 to have length δx and width $\ll \delta x$. The current density J_z (ampères/meter^2) can be introduced as an equivalent current i_z (ampères) throughout each boundary area. Our software solves the algebraic equations, which minimize the energy functional for combined excitation $\boldsymbol{J}$ and $\boldsymbol{J}_m$, yielding

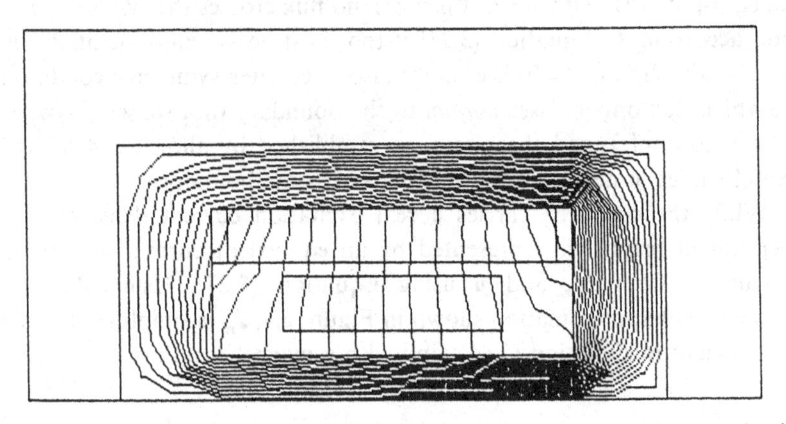

Figure 5.8. Flux distribution in a linear actuator with permanent magnet and coil excitation.

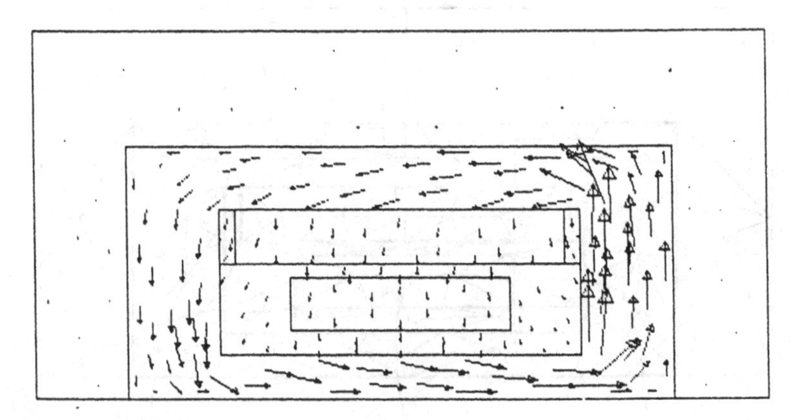

Figure 5.9. Flux density vectors in a linear actuator with permanent magnet and coil excitation.

a vector potential distribution throughout the region. Equi-potentials of A are coincident with flux lines, and can be used to illustrate the flux distribution as in Figure 5.8. By applying Equation (5.23) at each element, an alternative representation in Figure 5.9 shows flux density $\boldsymbol{B}$ vectors. This example illustrates the usefulness of performing a detailed analysis of the flux distribution in a device, either using the finite element method or some other technique. While the permanent magnets in this actuator "see" a symmetrical magnetic circuit about their axis and deliver their magnetic field equally towards both ends of the device, the addition of coil current causes the field to be concentrated towards one end. The plot of flux distribution cautions the designer to study possible saturation effects in those parts of the soft iron core.

References

Campbell, P., Chari, M. V. K. and D'Angelo, J. (1981). Three-dimensional finite element solution of permanent magnet machines. *IEEE Transactions on Magnetics*, **17**, 2997–9.

Campbell, P. and Al-Murshid, S. A. (1982). A model of anisotropic Alnico magnets for field computation. *IEEE Transactions on Magnetics*, **18**, 898–904.

Chari, M. V. K. and Silvester, P. P. (Ed) (1980). *Finite Elements in Electrical and Magnetic Field Problems*. New York: John Wiley & Sons.

Ramo, S., Whinnery, J. R. and Van Duzer, T. (1965). *Fields and Waves in Communication Electronics*. New York: John Wiley & Sons.

Smith, G. D. (1965). *Numerical Solution of Partial Differential Equations*. London: Oxford University Press.

6
Magnetizing and testing

6.1 Introduction

In previous Chapters, transfer of energy to and from a permanent magnet, and associated changes in energy of the external field have been discussed. For initial magnetization, the objective is to apply sufficient energy to the material to align its internal magnetization vectors $\boldsymbol{M}$ in a unique direction, for which the magnet is said to *saturate* at $\boldsymbol{M}_{sat}$. Optimum performance is achieved along this preferred axis, but the characteristics quantifying the material's properties have to be measured outside the magnet. The internal parameter $\boldsymbol{M}$ cannot in fact be measured directly, and although the *intrinsic* curve is the more fundamental characteristic of a magnet, it must be deduced from an external measurement of the normal B *versus* H loop.

This example illustrates a problem, which is commonly encountered by users of permanent magnets. Whereas they design a device to operate in a certain manner, it later appears that the real magnet does not meet these expectations. Later in this Chapter, we discuss various techniques that are used for measuring magnetic parameters – their application to the properties of magnets themselves will provide the basis for quality control. Unfortunately, the magnitude of applied field that is required to *saturate* a particular magnet is a somewhat empirical quantity.

6.2 Magnetization

It is generally considered that, to magnetize a fully dense permanent magnet to saturation, an external field of 3–5 times its intrinsic coercivity H_{ci} must be applied. Manufacturers will provide minimum magnetization requirements for their materials, which support this estimate. For example, Ceramic 8 having $H_{ci} = 250$ kA/m at room temperature will need an applied field of around 800 kA/m, and Alnico 5 with $H_{ci} = 50$ kA/m needs about 240 kA/m to saturate it. The disparity is because ceramic ferrites base their permanent magnetism on magnetocrystalline anisotropy, whereas the dominant mechanism in alnicos is shape anisotropy. By considering the energies associated with each of these anisotropies, as outlined in Sections 1.4 and 1.6, it is possible to calculate theoretical values for H_{ci} above which the magnetization saturates at M_{sat}. Applied field in excess of $|H_{ci}|$ is required for a *real* magnet, throughout the volume of which there will be significant variation in the *actual* composition and properties.

Many of the high energy rare earth magnets also require $\geqslant 3H_{ci}$ for magnetization, and because of their high intrinsic coercivities, these applied fields can become very large indeed. In Section 3.4, we discussed the effect of composition variations on the magnetic and thermal properties of rare earth magnets. Nd–Fe–B has an intrinsic coercivity of about 1000 kA/m, but when modified as (Nd, **Dy**)–Fe–B to improve the temperature coefficient, H_{ci} rises to about 1700 kA/m and the material becomes that much harder to magnetize. While further substitution of cobalt provides a desirable increase in Curie temperature, an important practical advantage of (Nd, **Dy**)–(Fe, **Co**)–B is the reduction of H_{ci} back to about 1000 kA/m and the correspondingly small applied field required for saturation.

Except for rapidly quenched "$Nd_2Fe_{14}B$", rare earth magnets do not observe the simple single domain model used for magnetocrystalline anisotropy. In fact, a variety of mechanisms control the coercivities and saturation requirements of the various compositions. The initial magnetization curve for rapidly quenched $Nd_2Fe_{14}B$ shown in Figure 2.19 illustrates the high field required to saturate even a *virgin* magnet. A similar characteristic was shown in Figure 2.15 for Sm_2Co_{17}, though coercivity in this material is controlled by *pinning* in its fine cell structure. However, it is the grain boundaries that provide pinning of the domain walls in *nucleation*-type magnets, such as sintered $Nd_2Fe_{14}B$ and $SmCo_5$. In *virgin* material of either type, the relatively easy domain wall motion within each grain allows saturation to be achieved with an applied field $\ll H_{ci}$, as was illustrated in Figure 2.12. There are therefore significant differences in magnetization requirements between $SmCo_5$ and Sm_2Co_{17}, and between sintered and rapidly quenched $Nd_2Fe_{14}B$.

Whereas domain wall motion governs the initial magnetization of nucleation-type $SmCo_5$ and $Nd_2Fe_{14}B$, this is not the mechanism that controls their subsequent remagnetization, when the material is no longer "virgin". Once the magnetization within a nucleation magnet has been aligned with an external field, any attempt to reverse its direction will be more a matter of applying $|H_{ci}|$ than of domain wall motion – intrinsic coercivity is defined as the field required to reverse M_{sat}. Consequently, the magnetization requirements for sintered $Nd_2Fe_{14}B$ and $SmCo_5$ are also dependent upon the history of the material. These differences are illustrated on the intrinsic magnetization loop shown in Figure 6.1. Starting with a virgin magnet that has never experienced an external field, saturation will be achieved along the initial curve (path ***a***) with an applied field $\ll +H_{ci}$. When brought into normal operation, the magnet follows the major loop into its second quadrant (path ***b***), but this condition may result from the application of an external demagnetizing field. If a field of *exactly* $-H_{ci}$ is applied, Figure 6.1 shows that the net $\boldsymbol{M}$ in the magnet becomes zero – half of the magnetization vectors are directly opposed to the other half. To fully remagnetize the material from the $-H_{ci}$ point, half of the vectors must reverse direction, which requires an applied field of approximately $+H_{ci}$ (path ***c***); this is much greater than the field to saturate a virgin magnet. Lastly, full remagnetization of a material that has been saturated in the opposite direction means that *all* the magnetization vectors must be reversed, which requires an applied field $\gg H_{ci}$ and probably approaching the roughly $3H_{ci}$ estimate (path ***d***).

The magnetization process involves storage of kinetic energy in a magnetic material, which in turn requires that work be done by an applied field. A permanent magnet will retain some of this kinetic energy upon removal of the applied field; it may later release some of this energy in establishing an external field of its own. Consider a material, which is

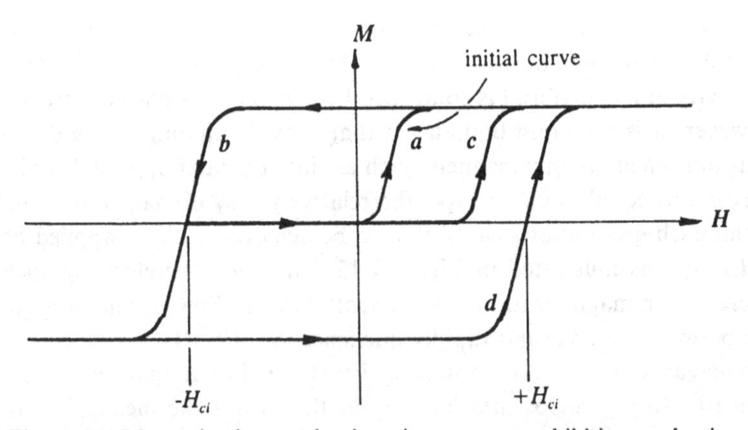

Figure 6.1. Magnetization mechanisms in a magnet exhibiting nucleation.

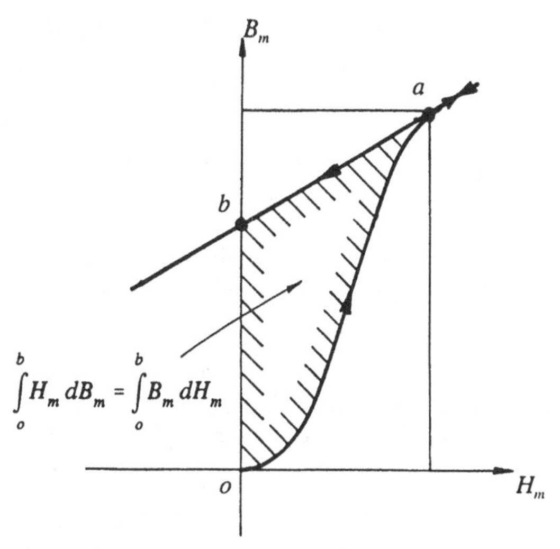

Figure 6.2. Change in energy per unit volume of a permanent magnet in its magnetization process.

magnetized from its virgin state (point ***o***) to saturation (point ***a***), as in Figure 6.2. Upon removal of the applied field, the magnet settles to its remanence (point ***b***), as it would if short-circuited by a soft iron "keeper". The total change in energy for each stage of this process was derived in Section 1.5, and may be written as

$$\int_o^a B_\mathrm{m}\,\mathrm{d}H_\mathrm{m} + \int_o^a H_\mathrm{m}\,\mathrm{d}B_\mathrm{m} = [B_\mathrm{m}H_\mathrm{m}]_o^a \tag{6.1}$$

$$-\int_a^b B_\mathrm{m}\,\mathrm{d}H_\mathrm{m} - \int_a^b H_\mathrm{m}\,\mathrm{d}B_\mathrm{m} = -[B_\mathrm{m}H_\mathrm{m}]_a^b \tag{6.2}$$

Per unit volume of magnet material, the second terms represent the internal kinetic energy stored, which is shown in Figure 6.2 to be the area enclosed by the path ***o–a–b***. Even if the applied field is increased beyond the level required for saturation at point ***a***, there will be no further increase in kinetic energy stored in the magnet. The first terms of Equations (6.1) and (6.2) are the work done by the applied field, which equals the kinetic energy stored around the complete path ***o–a–b***. This area gives a numerical value for the energy per unit volume required to saturate a magnet that is held in a keeper.

It is frequently not practical to achieve magnetization within the closed soft iron circuit of a keeper, and the applied field will then be required to overcome an air gap in order to saturate the permanent magnet. The operating point of the magnet is determined by the position of a *load line* as already defined in Equation (4.24), which incorporates the dimensions of the magnet and the air gap:

$$B_m = -\mu_0\left(\frac{k_1}{k_2}\right)\left(\frac{A_g l_m}{A_m l_g}\right)\left(H_m - \frac{Ni}{l_m}\right) \tag{4.24}$$

Lateral excursions of the load line along the H_m axis are dependent upon the coil excitation as shown in Figure 4.8; similar movement of the load line in the first quadrant is shown in Figure 6.3. Figure 6.2 is simply a special case of this with the magnet in a keeper, for which the air gap length $l_g = 0$, the slope of the load line is infinite, and all the coil excitation is directed into the magnet: $H_m = Ni/l_m$. However, with $l_g \neq 0$ and a finite load line slope, Figure 6.3 shows that $Ni/l_m > H_m$ and the coil excitation exceeds the magnetization in the permanent magnet by the amount required to overcome the air gap. Such an allowance must be made for the additional applied field required to overcome any gaps in the magnetizing circuit, or in the device itself if the magnet is to be magnetized "*in situ*".

In our discussion on dynamic operation of a magnet in Section 4.4, it was noted that, to a coil, a magnet appeared as its recoil permeability $\mu_0\mu_{rec}$. Because μ_{rec} has a value close to unity in all permanent magnets,

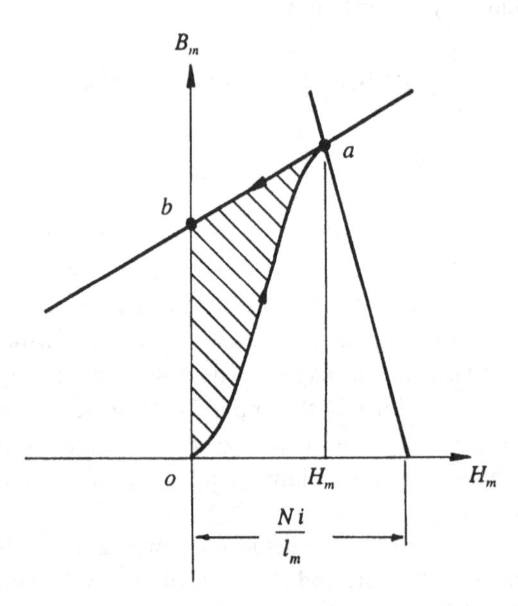

Figure 6.3. Magnetization of a permanent magnet in an air gap.

these materials do little themselves to guide a magnetizing field through them in the desired direction. Any shaping of the applied field must be achieved by careful design of the magnetizing fixture, a task made more difficult by the presence of additional air gaps in the circuit. An *anisotropic* material is processed in such a way that it exhibits superior magnetic properties along a preferred axis, but the saturating field must also be applied along this same axis for optimum performance to be realized. If magnetized in any other direction, the vector relationship between the magnet's parameters ($\boldsymbol{B}$, $\boldsymbol{H}$, $\boldsymbol{M}_r$ and $\boldsymbol{\mu}$) will prevail, as described by the permanent magnet model of Equation (5.44), and inferior performance is achieved with the magnet operating on a characteristic which falls within the major demagnetization curve. This has been illustrated for Alnico 5 by measuring the intrinsic characteristics at 10° increments from the preferred to the transverse axes, as shown in Figure 6.4 (Campbell and Al-Murshid, 1982). If a uniform, saturating field is not achieved throughout

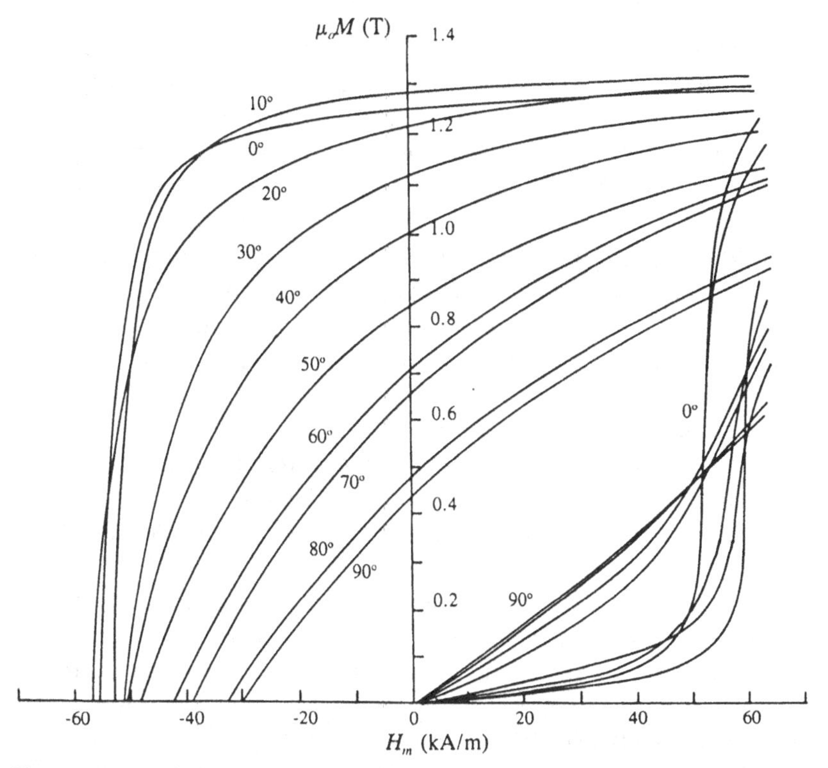

Figure 6.4. Intrinsic magnetization curves measured at various angles to the preferred direction in Alnico 5.

a magnet's entire volume, then its *effective* demagnetization characteristic will reflect that it has not been fully magnetized.

To fully magnetize a permanent magnet material, it is only necessary to achieve the saturating field within it momentarily. As greater fields are required to saturate the modern, higher coercivity materials, it becomes imperative to apply this field in a very short duration pulse to utilize the energy available from the magnetizer equipment. If the magnetization pulse has a fundamental frequency f, then it is well known that eddy currents will only allow this field to penetrate a metallic material to a depth δ, known as its *skin depth.* Clearly, we desire full penetration of the field both into the magnet and into any soft iron component attached to it, including the magnetizing fixture. This will depend upon the permeability $\mu_0\mu$ and resistivity ρ of each component (Bleaney and Bleaney, 1965), which determine the skin depth via

$$\delta = \left(\frac{\rho}{\pi f \mu_0 \mu} \right)^{1/2} \tag{6.3}$$

All permanent magnets have very low permeabilities, appearing substantially like air gaps to any surrounding circuit, and so resistivity plays the dominant role in the magnetization process. While a high $\rho > 10^{-4}\ \Omega$ m ensures full penetration of a pulse field into a ceramic ferrite magnet, values are much smaller in metallic magnets, for which skin depth may become important. Approximate resistivities are $150 \times 10^{-8}\ \Omega$ m for fully dense $Nd_2Fe_{14}B$, $85 \times 10^{-8}\ \Omega$ m for sintered Sm_2Co_{17}, falling to around $50 \times 10^{-8}\ \Omega$ m for $SmCo_5$ or alnico magnets. According to Equation (6.3), low ρ can be compensated by decreasing f, lengthening the pulse. Because it directly increases the energy drawn from the magnetizer, this is easier to implement for low energy alnicos than for high energy rare earth magnets. The most difficult design problem, however, is posed by the magnetizing fixture, where the combination of very low ρ with very high μ is frequently incompatible with delivery of high magnetization energy. A common solution for large rare earth magnets is to eliminate the fixture's soft iron core altogether and just use an *air-cored* coil.

Selection of the fixture's power supply depends upon the energy it must deliver to saturate the magnet. Direct current can be used for many ceramic ferrite and alnico materials, and it is even possible to use a high coercivity rare earth magnet as the field source to magnetize a much lower energy material. There are two common ways to generate the large pulsed fields needed to saturate higher energy magnets, which are used exclusively for $SmCo_5$, Sm_2Co_{17} and $Nd_2Fe_{14}B$.

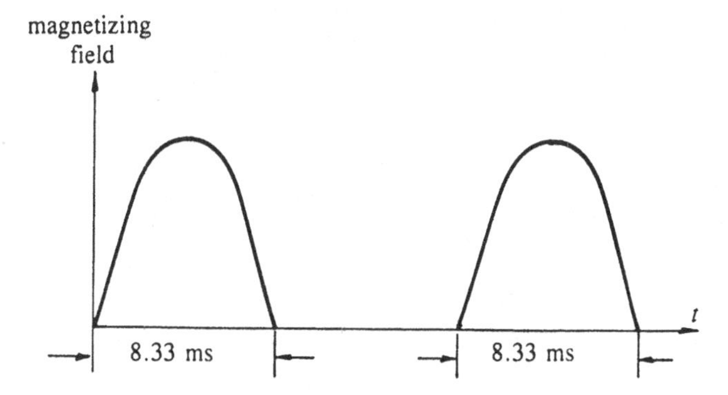

Figure 6.5. Current pulses from a 60 Hz half-cycle magnetizer.

In a half-cycle magnetizer, a.c. mains voltage is fed to the fixture either via an ignitron or via a thyristor, which at most will allow current during only half of each cycle to pass. In practice, a timing circuit is employed to vary the firing time during each half-cycle, thus shaping the current pulses. With 60 Hz mains frequency, the maximum duration for each pulse is 8.33 ms as shown in Figure 6.5. Because voltage is determined by the mains supply, a transformer is sometimes interposed between the fixture and the magnetizer circuit to increase the peak current delivered.

The most versatile magnetizer, and the most widely used today for all varieties of permanent magnets, is the capacitor discharge type. Energy is taken slowly from the mains supply to charge a bank of capacitors to a high d.c. voltage level, which are then discharged into the fixture as a single, short duration pulse via an ignitron (or a thyristor). The energy stored in the capacitor bank is $\frac{1}{2}CV^2$, which must not only be sufficient to saturate the magnet volume as indicated in Figure 6.2, but must overcome the inherent energy requirements of the magnetizing circuit, including fixture inductance ($\frac{1}{2}i^2L$) and resistive heating ($\frac{1}{2}i^2R$). The capacitors used to magnetize the largest rare earth magnets can be as great as 200 000 μF and may be charged to around 5000 V.

The basic elements of a capacitor discharge magnetizer are illustrated by the circuit in Figure 6.6. The capacitor C is charged to its initial voltage V_0 at a rate determined by the resistance R_0. When the ignitron is fired, V_0 is switched to the fixture, whose inductance L and resistance R should include the parasitic values in the magnetizer itself. After closing the switch, the dynamic behavior of the circuit is described by

$$L\frac{\mathrm{d}i}{\mathrm{d}t}+Ri+\frac{i}{C}\int_0^t i\,\mathrm{d}t=V_0 \tag{6.4}$$

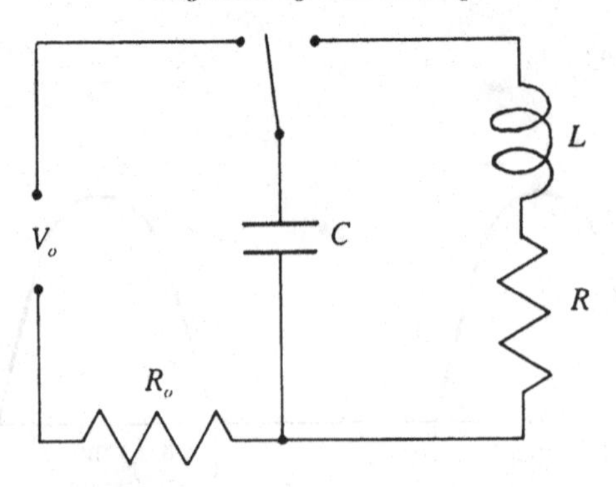

Figure 6.6. Electrical circuit of a capacitor discharge magnetizer and a fixture.

Differentiating this yields a conventional second-order differential equation:

$$\frac{\mathrm{d}^2 i}{\mathrm{d}t^2}+\frac{R}{L}\frac{\mathrm{d}i}{\mathrm{d}t}+\frac{i}{LC}=0 \tag{6.5}$$

Variations in the parameters, such as L changing due to the non-linear magnetization characteristic of its iron, require that a numerical method be employed to solve this equation (Lee, 1988). As an illustration, however, if linearity is assumed, then the current pulse will have the general form

$$i=K_1\,\mathrm{e}^{-t/\tau_1}+K_2\,\mathrm{e}^{-t/\tau_2} \tag{6.6}$$

The two time constants in this equation are found to be

$$\frac{1}{\tau_1}=\frac{R}{2L}+\left[\left(\frac{R}{2L}\right)^2-\frac{1}{LC}\right]^{1/2}$$
$$\frac{1}{\tau_2}=\frac{R}{2L}-\left[\left(\frac{R}{2L}\right)^2-\frac{1}{LC}\right]^{1/2} \tag{6.7}$$

The constants K_1 and K_2 are found by utilizing the initial condition at discharge, when at $t=0$ the capacitor voltage is V_0, the current $i=0$, and

$$\frac{\mathrm{d}i}{\mathrm{d}t}=\frac{V_0}{L} \tag{6.8}$$

Hence

$$K_2=-K_1=\frac{V_0}{2L\left[\left(\frac{R}{2L}\right)^2-\frac{1}{LC}\right]^{1/2}} \tag{6.9}$$

Initially, there is a rapid rise in current, which is dominated by the short time constant, τ_1. After peaking, the subsequent decay of current is controlled by the much longer time constant, τ_2. As indicated by Equations (6.6)–(6.9), the current pulse will be uni-directional without oscillations as shown in Figure 6.7, provided that

$$R > 2\left(\frac{L}{C}\right)^{1/2} \tag{6.10}$$

The energy contained in this plane can be calculated by utilizing $\int i\,dt$ over the duration of the pulse, but the total is just the energy released from the capacitor, $\frac{1}{2}CV^2$.

Equations (6.6)–(6.9) completely describe the operation of a capacitor discharge magnetizer under the assumption that the fixture has linear magnetic characteristics. A fixture should be designed so that its parameters (in R and L) are combined with the capacitor's (C and V_0) to ensure that sufficient energy is delivered to the target permanent magnet material, at a high enough current to produce a saturating applied field. This magnetizing field is directly related to the current via the winding turns N in the fixture, and follows the same form as the current pulse in Figure 6.7, the maximum value of which may be determined by differentiating Equation (6.6).

As already mentioned, the trend for magnetizing large, high coercivity rare earth magnets is to use an iron-less fixture, which only comprises an

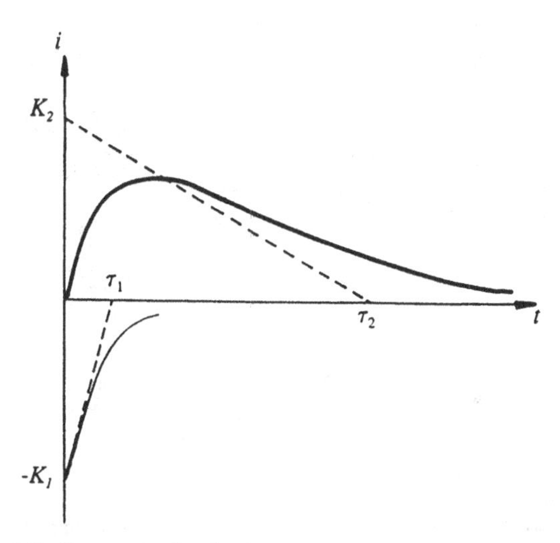

Figure 6.7. Current pulse from a capacitor discharge magnetizer.

Figure 6.8. Solenoic coil fixture for a capacitor discharge magnetizer. (Courtesy of Magnetic Instrumentation, Inc.)

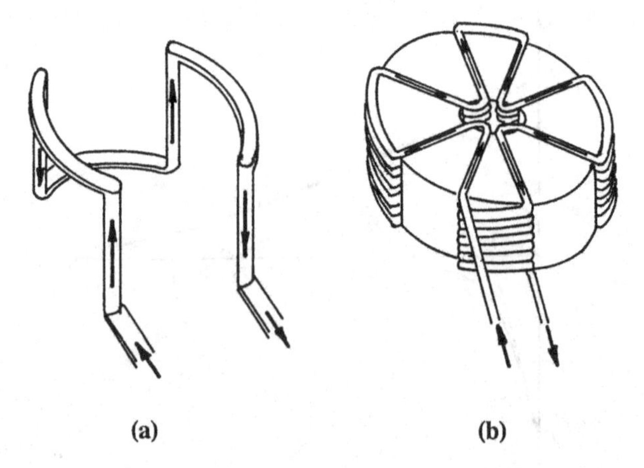

Figure 6.9. Magnetizing coils for multi-pole arrays of (*a*) four radial poles, and (*b*) eight axial poles.

air-cored coil. A single permanent magnet requiring uniaxial magnetization can be placed in the bore of a solenoid coil. If the length l_s of the solenoid is significantly greater than that of the magnet, l_m, then an essentially uniform magnetizing field is applied. The field in the bore of a "long" solenoid is well known to be axially directed and uniform, of magnitude

$$H_s = \frac{Ni}{l_s} \tag{6.11}$$

A typical solenoid fixture is shown in Figure 6.8, and larger coils of this type that deliver greater energy may need to be water-cooled. When more complex magnet shapes are involved, custom designed coils are generally required. Two examples are shown in Figure 6.9, for magnets with radial and axial multi-pole arrays, as employed in many permanent magnet electric motors.

6.3 Demagnetization

Users of permanent magnets occasionally have the need to *de*magnetize magnets, while magnet manufacturers frequently do so to facilitate handling and shipping. Several methods are available for demagnetization, but the choice will depend upon the intrinsic coercivity and size of the magnet. In fact, it is only necessary to apply a field of around $|H_{ci}|$, as was illustrated by path *c* in Figure 6.1. As a practical matter, it is quite difficult to apply the *exact* field, which, when removed, brings a magnet to $B_m = 0$ as shown in Figure 6.10. However, this technique is useful to make a high coercivity magnet amenable to complete demagnetization by another method.

The most common way to achieve demagnetization is to apply an alternating field, initially of sufficient magnitude to reverse the flux within the magnet (though not necessarily to saturate it in the opposite direction). The magnitude of this alternating field is then gradually reduced to zero, bringing the magnet flux down with it; alternatively, the magnet can be slowly removed from the field. For higher fields, a capacitor discharge magnetizer and coil can be used, but with the values of *R*, *L* and *C* chosen such that Equation (6.10) is not valid, and an oscillatory waveform of decaying magnitude is produced. The *skin depth* δ may prevent an alternating field fully penetrating the larger metallic magnets, in which case a d.c. supply may be used. Connections to this can be switched to

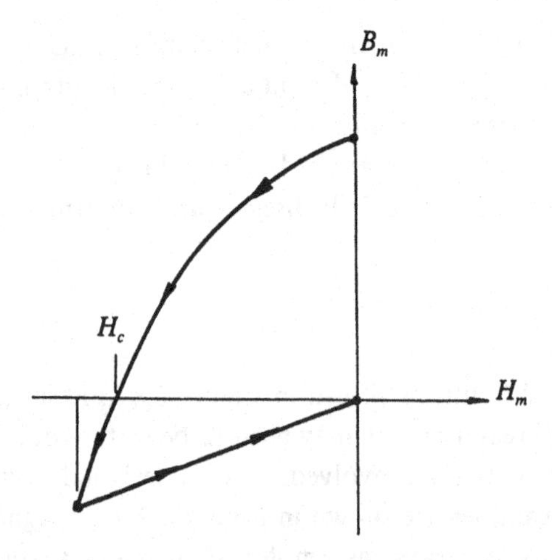

Figure 6.10. Demagnetization of a magnet by recoil.

apply the field alternately in opposite directions, while steadily reducing its magnitude.

Demagnetization of rare earth magnets can be assisted by application of heat, because their intrinsic coercivities decline with increasing temperature. Care must be taken, though, to avoid any metallurgical changes such as oxidation, and it is usual to assist alternating field demagnetization of rare earths with temperatures only up to around 200 °C. Ceramic ferrites, however, can be completely demagnetized thermally by exceeding their Curie temperature of 450 °C, above which their spontaneous magnetization is destroyed; this does not require the assistance of an applied field. It is important not to exceed about 1000 °C, though, above which *permanent* metallurgical changes in ferrites will occur. Thermal demagnetization does not work in alnicos, because *permanent* changes in their phase compositions appear at around 500 °C, well below their Curie temperatures.

6.4 Field measurement

When a coil experiences a changing magnetic field, there is an e.m.f. e induced in it, which may be used to measure the magnitude of the field itself. As introduced in Equation (4.39), *Faraday's Law* of electromagnetic induction defines e as the rate of change of flux linking the coil,

$$e = -\frac{d\lambda}{dt} \tag{6.12}$$

If the coil comprises N turns and the instantaneous magnitude of the flux is Φ, then

$$e = -N \frac{\mathrm{d}\Phi}{\mathrm{d}t} \tag{6.13}$$

A change in magnetic flux from Φ_1 to Φ_2 may therefore be measured by the e.m.f. in a search coil using

$$\int_{\Phi_2}^{\Phi_1} \mathrm{d}\Phi = \frac{1}{N} \int_0^t e \, \mathrm{d}t \tag{6.14}$$

If the coil's circuit is purely a resistance R, then its current i ($=e/R$) is also a measure of flux change via

$$\frac{N}{R}(\Phi_1 - \Phi_2) = \int_0^t i \, \mathrm{d}t = q \tag{6.15}$$

q is the total charge flowing in the circuit resulting from the flux change.

One instrument that uses this effect is the *ballistic galvanometer*, a moving coil meter, which is designed with very light damping. A permanent magnet provides a constant field through the coil, which is suspended in the gap between the magnet's poles and an iron core, as shown in Figure 6.11; the coil experiences a uniform field as it rotates. Current flowing in this coil interacts with the field, creating a torque, which causes it to rotate; this is balanced by a restoring torque due to the coil's suspension. In a ballistic galvanometer, the coil's inertia is sufficient that the entire charge q due to the flux change is dissipated in the coil before it begins to move. In effect, q gives the galvanometer coil an impulse, and the

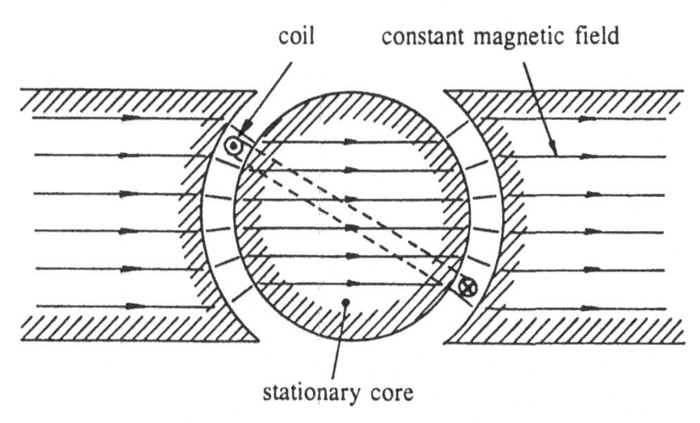

Figure 6.11. Moving coil in a permanent magnet field in a galvanometer.

subsequent motion is determined by the instrument's parameters, such as the inertia and the circuit resistance. These are chosen so that the motion is critically damped – the coil returns to its original position without oscillating in a minimum of time. To measure the magnitude of a steady flux, a search coil is connected to a ballistic galvanometer as shown in Figure 6.12; the search coil is moved quickly out of the field to a region of approximately zero flux, and the subsequent maximum deflection of the galvanometer coil is a direct measure of the flux change. The circuit is made less sensitive to the search coil's resistance by including resistances R_s, to adjust the sensitivity, and R_d, to control the damping; the galvanometer predominantly experiences R_d provided that $R_d \ll R_s$.

Another type of galvanometer that is used for measuring magnetic flux is simply known as a *fluxmeter*. Its mechanism is similar to that of the ballistic galvanometer, except that the suspension provides no restoring torque. When an induced e.m.f. due to a flux change in the search coil imparts an impulse to this galvanometer coil, current flows creating a torque. However, the subsequent motion is limited by electromagnetic damping – a back-e.m.f. due to the coil's motion in a strong permanent magnet field will bring the current to zero, and thus halt the motion of the galvanometer coil. The deflection of a fluxmeter coil follows the change of flux in the search coil, and does not return automatically to its original position; it must be returned to its zero position electrically before another measurement can be made. A fluxmeter is accurate to about 1%, whereas a properly calibrated ballistic galvanometer can achieve an accuracy of about 0.1%.

In practice, it is hard to build a fluxmeter in which the suspension does not provide at least a slight torque, the presence of which will cause a slow drift of the instrument's final reading. This effect may be compensated to some extent if the integration in Equation (6.14) is performed

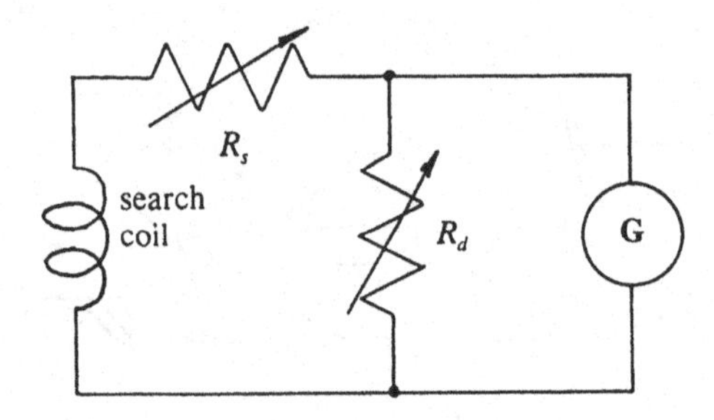

Figure 6.12. Circuit for critical damping of a galvanometer.

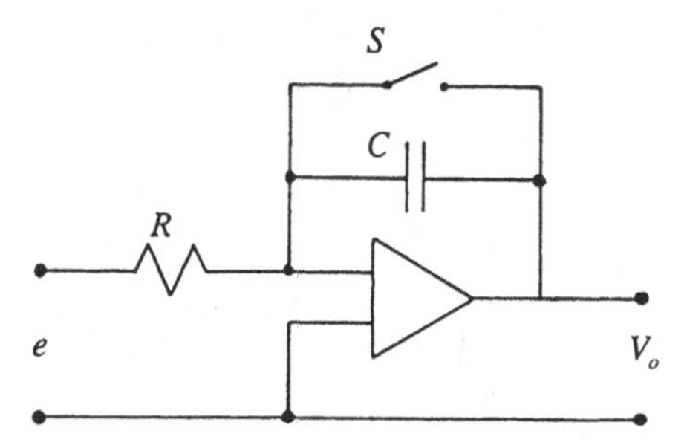

Figure 6.13. Integrating circuit using an operational amplifier.

electronically, instead of using a moving coil. In an electronic *integrating fluxmeter*, the e.m.f. e induced in a search coil is fed to an operational amplifier with capacitive feedback, as shown in Figure 6.13. Provided that the gain of the amplifier is very high, then the expression for its output voltage V_0 reduces to the simplified form

$$V_0 = -\frac{1}{RC}\int_0^t e\,\mathrm{d}t \tag{6.16}$$

Combining this with Equation (6.14), the flux change $\Delta\Phi$ measured by the search coil is represented as

$$V_0 = \frac{N}{RC}\Delta\Phi \tag{6.17}$$

With electronic integration, V_0 is not dependent upon the speed at which $\Delta\Phi$ is created, and the reading is maintained well after the search coil measurement is completed. In fact, the instrument reading must be returned to zero electronically by closing the switch S across the capacitive feedback. High gain in an amplifier is a recipe for voltage drift, but this is compensated in commercial integrating fluxmeters by adjusting bias currents in the circuit.

The various causes of drift in fluxmeters are not prevalent in solid state semiconductor devices, which have therefore become the most popular method for direct measurement of magnetic flux. Certain compounds such as gallium arsenide and indium antimonide exhibit a most useful characteristic known as the *Hall effect* in the presence of a magnetic field. An instrument using a sensor of this type is known as the *Hall effect gaussmeter*. The Hall effect is most simply described with reference to Figure 6.14, which shows the forces acting on a current-carrying electron e in a semiconductor element. An applied current i flows through the

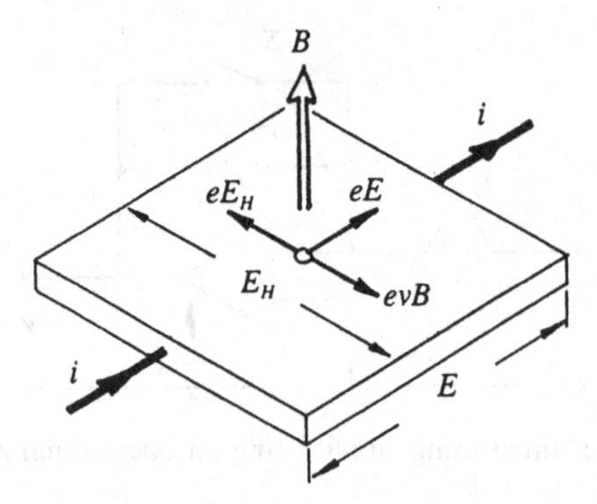

Figure 6.14. The Hall effect of forces acting on an electron moving in a semiconductor element.

element by virtue of an electric field E, which imparts a force of magnitude eE to each electron, giving it a velocity v. In the presence of a magnetic field B through the element as shown, there will be an additional component of force transversely across the element, of magnitude evB – the electron is deflected towards one side of the semiconductor. The build-up of electrons on one side causes an electrical charge, which, in turn, produces an additional electric field E_H and an associated force eE_H. Under equilibrium conditions, the transverse forces will balance so that the electrons can move freely along the element. The magnitude of the *Hall field* E_H is found by equating the transverse forces:

$$eE_H = evB \tag{6.18}$$

A density n of conduction electrons moving at an average velocity v will cause a current density through the element $J = nev$, so

$$E_H = \left(\frac{1}{ne}\right) JB = R_H JB \tag{6.19}$$

This provides a definition for the *Hall constant* R_H, and shows that the Hall voltage V_H developed across the width w of an element is proportional to the current i along its length, and the flux density B through its thickness t. Noting that $E_H = V_H/w$ and that $J = i/wt$, the Hall voltage is found from Equation (6.19) to be

$$V_H = \left(\frac{R_H}{t}\right) iB \tag{6.20}$$

Equation (6.20) describes the ability of a Hall element to measure a magnetic field B. The sensitivity can clearly be increased by raising the supply current, though this approach is of limited value because the consequent heating changes the semiconductor characteristics, including the Hall constant R_H. A greater thickness to the element provides a more robust device, but also reduces the sensitivity. A Hall element's sensitivity is fundamentally determined by the electron mobility. Indium antimonide (InSb) has a high mobility and is therefore a very effective Hall element material, but its R_H varies significantly with temperature as illustrated in Figure 6.15. While gallium arsenide (GaAs) has a much lower mobility and lower sensitivity than InSb, it is much more stable over a wide temperature range. A *Hall effect gaussmeter* using a Hall element and appropriate signal processing circuitry will measure flux density with high accuracy, stability and linearity; because the Hall effect is independent of frequency, this measurement is provided from d.c. to microwave frequencies.

A Hall sensor provides a convenient means for determining the flux density B passing through it in an air gap, but it cannot be used for flux measurement *within* a magnetic material – providing a slot for access by the sensor itself creates an air gap disturbing the internal field. The total flux through a magnetic material can, however, be measured using a search coil, which fits closely over it, connected to a fluxmeter or, more commonly, an integrating fluxmeter; knowing the coil's area then yields the internal flux density B. Magnetizing force in an air gap is directly proportional to B via $H=B/\mu_0$ (Equation (4.14)). H cannot be deduced from a measurement of B within either a soft or permanent magnetic material, however, because their relationship is more complex:

$$\boldsymbol{B}=\mu_0[(1+\chi)\boldsymbol{H}+\boldsymbol{M}_r] \tag{5.6}$$

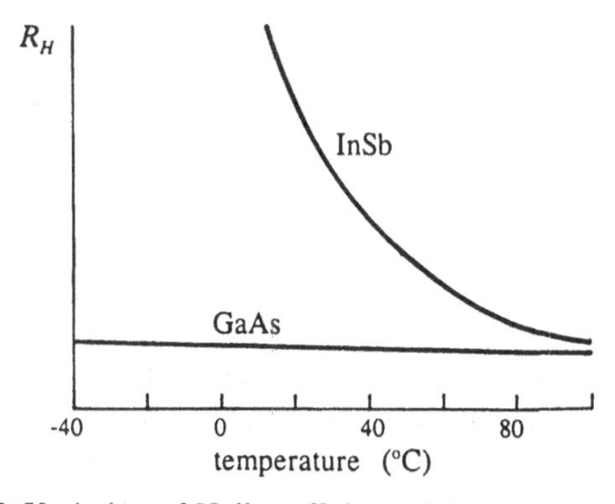

Figure 6.15. Variation of Hall coefficient with ambient temperature.

This is true even if B is found using a search coil around the magnetic material.

The magnetizing force H within a magnetic material must therefore be deduced by indirect measurements made outside the material. In Equation (5.20), we showed that the *tangential* component of $\boldsymbol{H}$ is continuous across a magnetic interface only when surface currents are not present. This is the case with soft magnetic materials, and so if $\boldsymbol{H}$ within the material is parallel to its surface, it will equal the value of $\boldsymbol{H}$ parallel to and just outside the interface; this can be directly deduced from the flux through a properly oriented Hall sensor or small search coil placed against the material. Permanent magnets, however, have a remanent magnetization $\boldsymbol{M}_r$, which is modeled by *equivalent* surface currents via Equation (5.47), and was depicted in Figure 1.4. Tangential $\boldsymbol{H}$ will not be continuous across a magnet's boundary, and measurement of the external field close to its surface will yield an erroneous result. Away from this localized effect at the surface, it is usually possible to find a closed loop through which no real current passes, and then to utilize

$$\oint \boldsymbol{H} \cdot \mathrm{d}\boldsymbol{l} = 0 \tag{4.8}$$

The most common way to establish this condition is to install the permanent magnet between the pole pieces of an electromagnet, whose excitation may then be used to apply a uniform field across the sample.

Figure 6.16 shows a magnet sample between the poles of an electromagnet, in which the magnet's cross-sectional area is considerably smaller than the surface area of the pole tips. This allows the electromagnet to deliver an approximately uniform field across the gap, including the sample, provided that its yoke is made from a high permeability soft iron material to ensure that the poles act as equipotential surfaces. If Equation (4.8) is applied around a closed loop that includes paths $\boldsymbol{a}$ and $\boldsymbol{b}$ shown in Figure 6.16, there will be no significant contribution to the integral from the paths through each pole tip joining $\boldsymbol{a}$ and $\boldsymbol{b}$. Furthermore, if the pole surfaces are co-planar, then the lengths of paths $\boldsymbol{a}$ and $\boldsymbol{b}$ are identical, and H within the magnet is equal to H elsewhere in the uniform field in the gap. As illustrated in Figure 6.16, H in the magnet is deduced from a measurement of B $(=\mu_0 H)$ along path $\boldsymbol{b}$ using either a Hall sensor or a search coil. One other method that is depicted is the use of search coils for B or H, which are embedded in the pole tips. This utilizes Equation (5.19), by which it was shown that the *normal* component of $\boldsymbol{B}$ is continuous across a magnetic interface. Thus, provided that the uniform field across the gap is normal to the co-planar pole surfaces, measurements using embedded search coils will reflect the values of B across the interface: B in the magnet directly, and H in the magnet from B in the adjacent gap.

Design of a permanent magnet requires its magnetization characteristic, a plot of which is obtained with the apparatus of Figure 6.16 by varying the electromagnet excitation while measuring B and H. The direct relationship between B, H and M (Equation (1.26)) allows M to be deduced and the intrinsic characteristic to be plotted concurrently. The characteristics may be plotted directly on a recorder, and a complete testing system of this kind is commonly called a *hysteresigraph*. Measurements made using search coils are of flux Φ, and the area of the coil must be known to convert this to flux density B. An embedded search coil under the sample magnet must be small relative to the magnet's area and in a narrow slot, to ensure that there is minimal distortion of the flux crossing the interface. With this proviso, the use of embedded coils is most convenient in hysteresigraphs because their areas are fixed, and the readings are independent of the size of the sample.

High energy rare earth permanent magnets present a unique difficulty in a hysteresigraph, because they can be excited to such high field levels

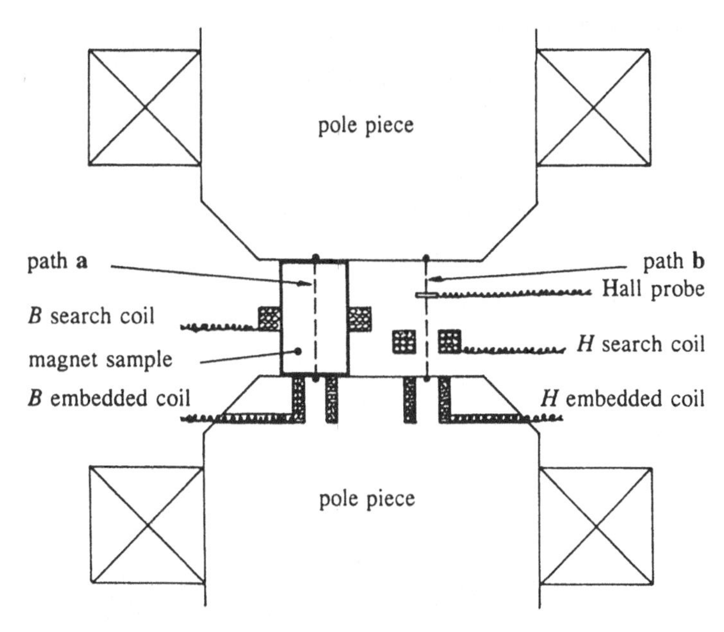

Figure 6.16. Magnet sample between the poles of an electromagnet, and various methods of measuring its B and H.

that the adjacent pole tips may become locally saturated. Under this condition, the poles are no longer true equipotentials, and the foregoing methods may produce erroneous measurements of H. This problem is overcome by using a *potential coil*, which utilizes the definition of $\boldsymbol{H}$ as the gradient of a scalar potential Ψ (Equation (5.17)). Consider the long, curved coil in Figure 6.17, which comprises a large number of turns of very small diameter, equally distributed along a non-magnetic core. When placed in a magnetic field where there are no real currents present, Equation (4.8) may be applied to the closed loop formed by path $\boldsymbol{a}$ along the coil's axis and path $\boldsymbol{b}$ outside the coil. The potential difference between the two ends of the coil may then be expressed as

$$\Psi_1 - \Psi_2 = \int_a \boldsymbol{H} \cdot \mathrm{d}\boldsymbol{l} = \int_b \boldsymbol{H} \cdot \mathrm{d}\boldsymbol{l} \qquad (6.21)$$

At any location on path $\boldsymbol{a}$, the field $\boldsymbol{H}$ along the coil (in the direction $\mathrm{d}\boldsymbol{l}$) is equivalent to a flux $\mu_0\, \delta A\, \boldsymbol{H}$ through the coil's area δA. If there are N_u turns per unit length, then the total flux linkage with the coil is

$$\lambda = \mu_0 N_u\, \delta A \int_a \boldsymbol{H} \cdot \mathrm{d}\boldsymbol{l} \qquad (6.22)$$

Substituting from Equation (6.21) yields

$$\lambda = \mu_0 N_u\, \delta A\, (\Psi_1 - \Psi_2) \qquad (6.23)$$

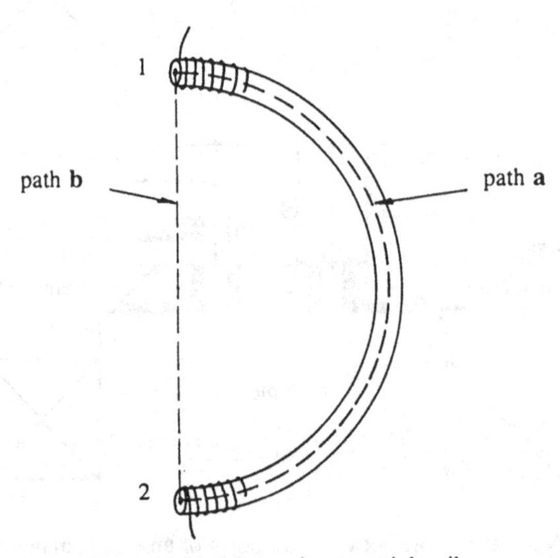

Figure 6.17. Magnetic potential coil.

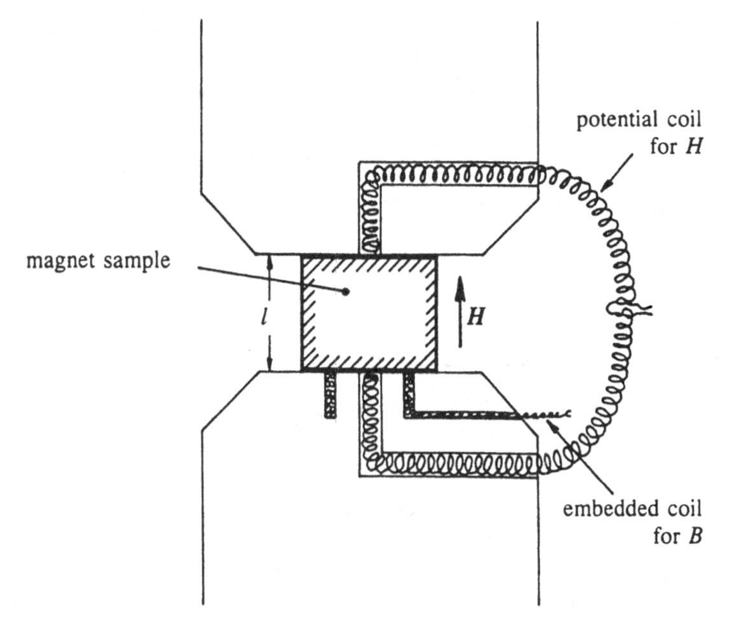

Figure 6.18. Potential coil measuring the potential difference between the ends of a magnet sample in an electromagnet.

The flux linkage λ in a potential coil is measured as with other types of search coil, according to Equation (6.12). However, λ is proportional to the potential difference between the *ends* of the potential coil, irrespective of its shape – one just needs to know the coil parameters N_u and δA. A potential coil can therefore be installed within a narrow hole in the yoke of an electromagnet as shown in Figure 6.18, with its ends at the opposite pole faces. The flux linkage measured using the coil yields the potential difference directly across the magnet sample, and according to Equations (6.21) and (6.23), this gives $\int \boldsymbol{H}\cdot d\boldsymbol{l}$ along the length l of the sample. Figure 6.18 also shows an embedded search coil for simultaneous measurement of B in the sample.

In a hysteresigraph, a magnet is captured in a short-circuit condition, and the field applied to it is varied using the electromagnet's coils. The measurement techniques may also be applied to an actual device, to determine the operating point of a permanent magnet under load. Consider the simple magnetic circuit of Figure 4.4, in which the magnet is operating on a load line of slope $-S$. A potential coil is installed as shown in Figure 6.19; with one end fixed in position $\boldsymbol{c}$, a fluxmeter reading is first taken with its remote end placed at one end of the magnet (point $\boldsymbol{a}$), and then

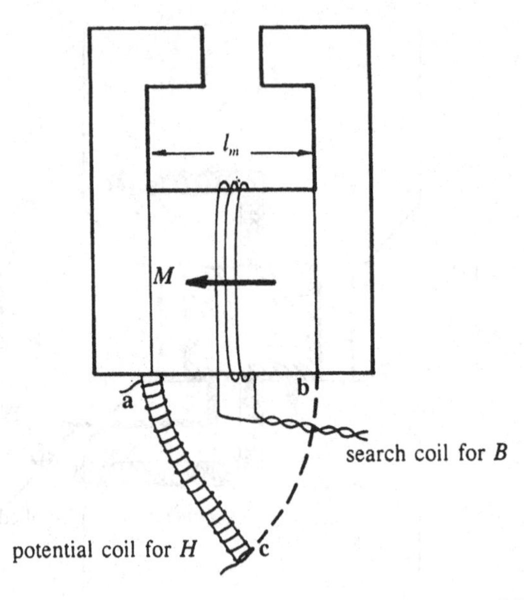

Figure 6.19. Measurement of B_m and H_m in a permanent magnet within a magnetic circuit with an air gap.

another reading is taken with it moved to the other end of the magnet (point ***b***), giving the potentials at points ***a*** and ***b*** relative to ***c***. The difference in the readings is the potential difference across the length of the magnet l_m, giving its magnetizing force H_m. The magnet's flux density B_m is deduced from the flux linking a close-fitting search coil, connected to a fluxmeter or integrating fluxmeter; the absolute value is obtained relative to a reading in no magnetic field, this change being achieved by removing the search coil from the device. The magnet's operating load line is then found to be $-S = B_m/\mu_0 H_m$. Clearly, the potential coil can be used to determine H in other lengths of the magnetic circuit, which may be useful in assessing the m.m.f. drop in various components. Likewise, strategic placement of search coils allows a detailed analysis of the flux distribution and leakage in a device, though removing each coil for an absolute reading can be tedious.

For high coercivity magnets, which retain their full magnetization even on open-circuit, the external field that they produce can be used to check their magnetic characteristic, and this is a convenient method for controlling the quality of large numbers of similar magnets in production. If the magnet's flux density B_m can be found from a search coil measurement, and its open-circuit load line is known, then a point on the demagnetization characteristic can be plotted to determine whether the expected magnetization is achieved in the sample. Even if it is not convenient to use a close-fitting search coil to link the entire magnet flux, it may be sufficient to use a *relative* measurement, such as dropping the magnet through the bore of a solenoid and recording $\int e \, dt$ (Equation (6.14)). An alternative to the solenoid is a pair of Helmholtz coils (Zijlstra, 1967), which allow better accessibility to the sample while providing a relatively large internal cavity where a uniform field produces a constant signal.

References

Bleaney, B. I. and Bleaney, B. (1965). *Electricity and Magnetism*, 2nd edn. Oxford: The Clarendon Press.

Campbell, P. and Al-Murshid, S. A. (1982). A model of anisotropic Alnico magnets for field computation. *IEEE Transactions on Magnetics*, **18**, 898–904.

Lee, J. K. (1988). The analysis of a magnetizing fixture for a multipole Nd–Fe–B magnet. *IEEE Transactions on Magnetics*, **34**, 2166–71.

Zijlstra, H. (1967). *Experimental Methods in Magnetism*, Amsterdam: North-Holland Publishing Co.

7
Applications

7.1 Introduction

Permanent magnets have been employed in a wide range of electrical apparatus for a great many years, and it is well beyond the scope of this text to discuss their design for all current applications. However, the dramatic improvements in material properties that accompanied the evolution of rare earth magnets have focussed interest on certain electromechanical and electronic devices, in which these materials may be applied to advantage. With this in mind, we discuss a wide variety of applications, which reflect the scope of new design activity today. This includes devices whose extremely high production quantities continue to make low cost ceramic ferrite the dominant material in today's market, and high added value products, which exhibit significant performance benefits using high energy rare earth magnets.

The most important application for permanent magnet materials is in direct current (d.c.) rotating electric motors. Ceramic ferrites have long been used in these machines to provide a steady magnetic field from their stators, but more recently rare earth magnets have been employed to particular advantage to promote the evolution of electronically commutated *brushless* d.c. motors, in which the permanent magnet assembly usually becomes the rotating component. The high energy of rare earth magnets is often used to produce a greater air gap flux density in a d.c. motor, which yields a corresponding improvement in the motor's output torque. The magnet's high coercivity is also attractive, because this improves its resistance to demagnetization from the motor's own armature winding. A variety of new motor topologies have evolved to meet more demanding packaging requirements, made possible by the properties now available from the extended range of modern permanent magnet materials.

A major market for permanent magnet d.c. motors is in automobiles, where cost usually overrides any performance considerations, and ceramic ferrites are almost universally employed. Modern automobiles embody an increasing number of electrically controlled functions, and various types of $Nd_2Fe_{14}B$ have indeed been investigated for use in improved d.c. motors for blowers, radiator cooling fans, starters and other applications. However, automobiles frequently present a hostile environment to these magnets, including severely corrosive elements and operating temperatures approaching 175 °C. It is the more docile environment within most consumer products that allowed brushless d.c. motors using either samarium–cobalt or neodymium–iron–boron magnets to become established, the most notable application being in computer hard disk drives. The development trend in hard disk drives is ever towards greater miniaturization, which led to the early use of Sm_2Co_{17} and then $Nd_2Fe_{14}B$ in brushless d.c. *spindle* motors – driving the spindle, which carries the rigid disk platters. Many other consumer appliances use permanent magnet d.c. motors, particularly if they are battery powered. The premium price of rare earth magnets over conventional types such as ceramic ferrite has to be justified by some benefit to the product, such as improvement in its performance, layout or ergonomics, or perhaps simplifying its assembly (as by using a bonded magnet material).

Direct current electric motors, whether mechanically or electronically commutated, are but one class of rotating electrical machine employing permanent magnets. Another major category is stepper motors, or *steppers*, which are particularly amenable to digital control because they rotate in a sequence of discrete steps. In less widespread use are synchronous machines, in which the permanent magnet and its field rotate in synchronism with a multi-phase stator winding field; a special case of this with similar construction is known as the hysteresis motor.

There is a wide variety of electromechanical devices that are known generally as *actuators*, which may or may not employ permanent magnets. Converting electrical energy into mechanical energy, they may either provide torque in a rotary system, or linear force. Rotary actuators are very similar or even identical to the various types of electric motor already mentioned – they simply have a limited range of angular movement. Linear actuators also frequently utilize conventional electric motors, whose rotary outputs are translated to linear motion through appropriate mechanical gearing. There are various means of coupling a permanent magnet's field with electrical and/or mechanical elements to directly produce linear force, one popular application being the hammer mechanism used in printers.

The importance of computer hard disk drives as a market for rare earth magnets has already been mentioned. A moving coil actuator is used to position the read/write head in a disk drive, and the demanding requirements both of miniaturization and of improved performance have made this a major application for fully dense $Nd_2Fe_{14}B$. Whereas the spindle motor runs at a constant speed, the head actuator must develop a high force to minimize the time taken to access information on the rigid disk platters. The head actuator magnet is therefore much larger than the spindle motor magnet, and it is the combination of its energy density and unit cost that has led $Nd_2Fe_{14}B$ to dominate this application in hard disk drives.

Moving coil actuators of this type are frequently known as *voice coil* actuators, because of their similarity to audio speaker systems. Audio speakers have traditionally employed alnico or ceramic ferrite permanent magnets, but the higher energies of rare earth materials afford a reduction in size and weight, which is attractive in a wide range of portable devices. It was the portable cassette player that provided the first significant market for $SmCo_5$ in the early 1980s, allowing miniaturization both of the headphones and of the capstan drive motor. In the 1990s, $Nd_2Fe_{14}B$ is increasingly used in automobile audio speakers, where the multiple speakers employed in a vehicle yield a worthwhile net saving in weight.

A specialized but important application for permanent magnet materials is in *sensors*, and while most require only relatively small magnets, some markets involve very high production quantities. Magnetic position and speed sensors are used in a variety of automobile system controls, because they provide a reliable, rugged, non-contact method of sensing in an often hostile environment. An example is measurement of crankshaft position, which is performed almost exclusively by magnetic sensors. While a sensor does not require a particularly large magnet to operate, the high energy densities of the rare earth materials are used to advantage either to increase the sensitivity of the device, or to allow greater running clearances (and more generous assembly tolerances) to be employed.

While the foregoing examples of permanent magnet applications have been selected for their importance on the basis of large production volumes, there are other examples where quite small quantities are offset by the very large size of magnet that is used. Two such applications are in magnetic resonance imaging systems, and for magnetic levitation of urban transit vehicles, usually at low speeds. Ceramic ferrites and $Nd_2Fe_{14}B$ are employed in products of each type, and in both cases they compete with superconducting magnets as the field source.

7.2 d.c. motors

The most significant application for most types of permanent magnet material is in direct current (d.c.) rotating electric machines. An armature winding supplied from a d.c. power source provides a simple means of controlling a motor's output speed. A d.c. machine may also employ a d.c. field winding, in series or in parallel with the armature winding, but a permanent magnet frequently offers a more efficient and less complex alternative to producing the main magnetic field. The most common type of permanent magnet d.c. machine is the electric motor, in which electrical energy is converted into mechanical torque and speed. Conversely, the rotor can be driven by a prime mover, and the machine functions as a d.c. generator producing electrical power. The description here concentrates on the *d.c. motor*, taking this to mean a d.c. rotating electric motor with a permanent magnet field system, and any major differences in the operation of a permanent magnet d.c. generator will be noted.

The range of performance features offered by d.c. motors broadened significantly with the coming of high energy rare earth magnets. Being able to produce the main field with a much smaller and lighter permanent magnet opened up new markets, and gave birth to the *brushless* d.c. motor. In conventional d.c. machines, the direction of current in the armature winding is switched mechanically through brushes bearing upon a commutator (Say and Taylor, 1980), whereas in a brushless d.c. machine, this switching is performed electronically (Miller, 1989). Electronic switching generally dictates that the armature winding be stationary, rather than rotating as in mechanically commutated machines, and the permanent magnet structure is part of the rotor rather than the stator. This difference is illustrated in Figure 7.1.

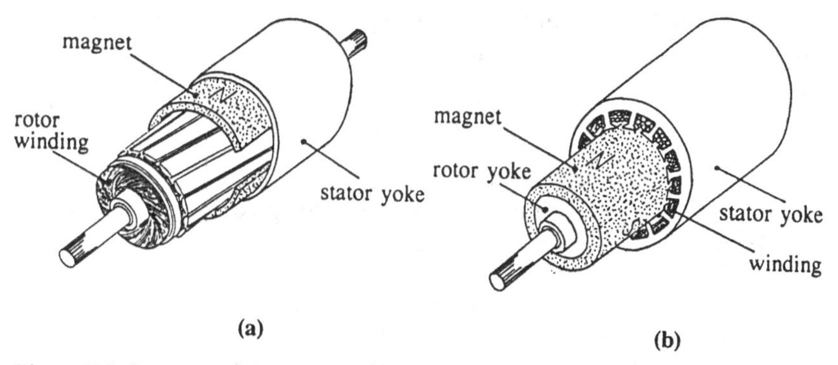

Figure 7.1. Layouts of the rotor and stator structures in (*a*) a permanent magnet d.c. commutator motor, and (*b*) a brushless d.c. motor.

In the conventional d.c. commutator motor shown in Figure 7.1(*a*), the armature winding is part of the rotor structure, rotating within the permanent magnet stator assembly. As the winding rotates, the current in it is switched through brushes in the stator and a commutator attached to the winding (not shown here, but shown later in Figure 7.8). The purpose of this switching is to maintain a consistent orientation between the permanent magnet field and the armature currents, regardless of the angular position of the rotor. In the brushless d.c. configuration of Figure 7.1(*b*), the armature winding is usually in the stator structure, within which is the permanent magnet rotor. This inverted topology is most common for brushless d.c. motors, although not mandatory; it does, however, provide the most convenient access to the winding terminals, and more effective removal of winding heat. The angular position of the magnet poles is usually detected magnetically with Hall effect sensors (not shown), and their signals are used to time the switching of the winding current, again to maintain a consistent orientation with the magnet's field. The high energy of rare earth magnets may be used to enhance the performance of d.c. motors in a variety of ways; it is, for example, possible to construct a rotor with very low inertia in a brushless d.c. motor.

Permanent magnet motors have some inherently attractive characteristics, which may be understood with reference to a few basic relationships. *Faraday's Law* of electromagnetic induction was used to derive the e.m.f. induced by a flux change in a simple coil as Equation (6.13). Now, an electric motor will utilize a heteropolar magnet structure where p is the number of poles (an even number), and each pole produces a flux of magnitude Φ. The armature winding will comprise a total of N turns, but these will generally be divided equally between a number of parallel paths a between its terminals. If the angular velocity of the rotor is Ω (radians/second), then Faraday's Law applied to any type of d.c. machine can be derived from Equation (6.13) as

$$e=\left(\frac{p\Phi N}{\pi a}\right)\Omega \tag{7.1}$$

If Φ is approximately constant, as it is in most permanent magnet devices, then the induced e.m.f. is directly proportional to speed. It is then common to define an *e.m.f. constant* of the machine as

$$K_{\mathrm{E}}=\frac{p\Phi N}{\pi a} \tag{7.2}$$

For a d.c. motor operating under steady state conditions at a supply voltage V and current i, the winding will appear to be purely resistive with resistance R, and the electric circuit equation is

$$V = e + iR \tag{7.3}$$

In a d.c. generator, terminal voltage is a result of the e.m.f. generated by driving the machine, and the circuit equation changes to

$$V = e - iR \tag{7.4}$$

In either case, under dynamic operating conditions, it will be necessary to include the winding's inductance L via an additional $L\,\mathrm{d}i/\mathrm{d}t$ term.

Combining Equations (7.1)–(7.3) shows that, with an almost constant field, the speed of a permanent magnet d.c. motor will only vary with current if the supply voltage is fixed:

$$K_E \Omega = V - iR \tag{7.5}$$

By a suitable choice of parameters, a fairly narrow speed range can be maintained for a wide variation in current, as shown in Figure 7.2. For this reason, permanent magnet d.c. motors can often operate open loop in speed control situations, and even where closed loop control is required, this characteristic makes control relatively easy to implement. A conversion from electrical power to mechanical power occurs at a motor's armature winding. After subtracting the i^2R heating from the power input Vi, ei is

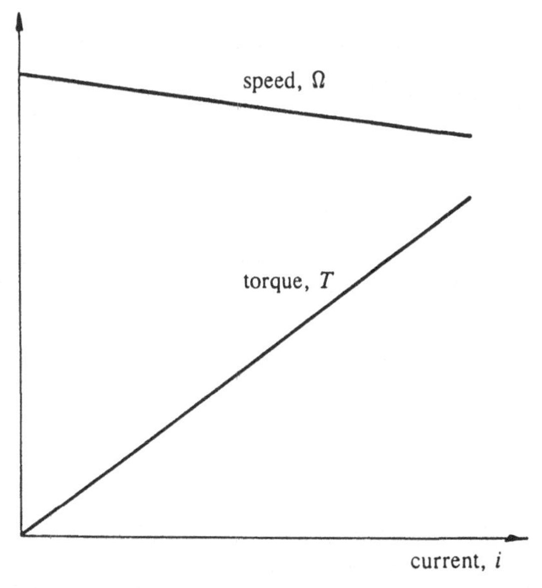

Figure 7.2. Speed and torque characteristics of a permanent magnet d.c. motor.

available for conversion into rotation with an "electromagnetic" torque T via

$$ei = \Omega T \tag{7.6}$$

The motor's output torque will actually be slightly less than T due to internal friction losses, represented by a no-load torque T_{nl}, but this difference will be neglected for clarity in this analysis. Combining Equations (7.1) and (7.6) shows that torque is directly proportional to current, as shown in Figure 7.2.

$$T = K_T i \tag{7.7}$$

In this case, it is more usual to use a *torque constant* of the machine, K_T, which is clearly equal to K_E as given in Equation (7.2). The units of K_T are N m A^{-1}, directly equivalent to K_E expressed in volt seconds/radian; however, because the speed of a machine is frequently calculated in revolutions per minute or revolutions per second, several different definitions for K_E are used in motor design texts (Kenjo and Nagamori, 1985). The main drawback to the operating characteristics of a permanent magnet d.c. motor, as shown in Figure 7.2, is that it possesses a very large *stall* torque, which may occur at an unacceptably high current of V/R (from Equation (7.3) at $\Omega = 0$).

In a cylindrical d.c. motor of either type shown in Figure 7.1, the running clearance between the motor and stator is made quite small to optimize the magnetic circuit performance. The length L and diameter D of the active air gap are therefore taken to be those of the rotor. The air gap flux density B_g may then be related to the flux per pole Φ via

$$B_g = \left(\frac{p}{\pi DL}\right)\Phi \tag{7.8}$$

Just as B_g gives a measure of the density of magnetic field in the useful air gap, so is it common to specify a *specific electric loading* A_c for an electric motor as the density of electric current at the armature surface – actually, the number of Ampère conductors per unit circumference of the surface:

$$A_c = \left(\frac{Z}{a\pi D}\right)i \tag{7.9}$$

The armature winding comprises a total of Z conductors, twice the number of turns N. Using $K_T = K_E$ from Equation (7.2), the torque in Equation (7.7) may be rewritten using Equations (7.8) and (7.9) in a form allowing the effects of different magnetic properties to be more clearly understood:

$$T=\left(\frac{\pi D^2 L}{2}\right)B_g A_c \tag{7.10}$$

In the design of a d.c. motor, there is clearly a trade-off between the air gap flux density B_g, which depends on the permanent magnet material that is used, and the specific electric loading A_c, which is related to the current density in the armature and to thermal considerations. A conventional armature is made from soft iron, with slots in the surface to accommodate the winding conductors – the interspersed teeth must carry the magnetic flux to the air gap. This structure is illustrated in Figure 7.1, in the rotor of (*a*) and in the stator of (*b*). A higher flux density requires a greater cross-section in the armature teeth to avoid saturation, whereas a larger electric loading means an increased slot area to house the winding and thus a reduced cross-section in the teeth.

If the dimensions of a d.c. motor are fixed, then the rotor will have a given size factor $\pi D^2 L$. Equation (7.10) then shows that the torque developed is increased by raising the air gap flux density, which is achieved by employing higher energy density permanent magnet materials. Significant applications for d.c. motors using fully dense Sm_2Co_{17} or $Nd_2Fe_{14}B$ have been found, where increased torque is needed (as in *electric* power steering in an automobile), or where torque is required with minimum size and weight, as in the aerospace environment. Improving torque by this means is preferable to increasing the electric loading, which would also raise the heating power losses. A higher air gap field is *always* the key to improving the performance of any permanent magnet device.

With an emphasis on achieving higher flux density levels in a d.c. motor to improve its performance, high remanence magnets would seem to be preferred. However, in an electric motor, the reaction flux from the armature winding interacts with the main flux due to the permanent magnets in the manner shown in Figure 7.3. If the magnets are adjacent to the air gap, this *armature reaction* will cause different operating conditions to exist in different regions of the magnets. At the trailing edge of each magnet, the operating point will be driven down the demagnetization curve in the manner described with reference to Figure 4.12. This dynamic condition may cause an irreversible loss of magnet flux in this region of every pole. At the leading edge of each magnet, however, armature reaction reinforces the main flux – a reversible effect. In a d.c. motor under load, the air gap flux density B_g will not be uniform under each pole, and the armature teeth at the trailing edges may be liable to saturate.

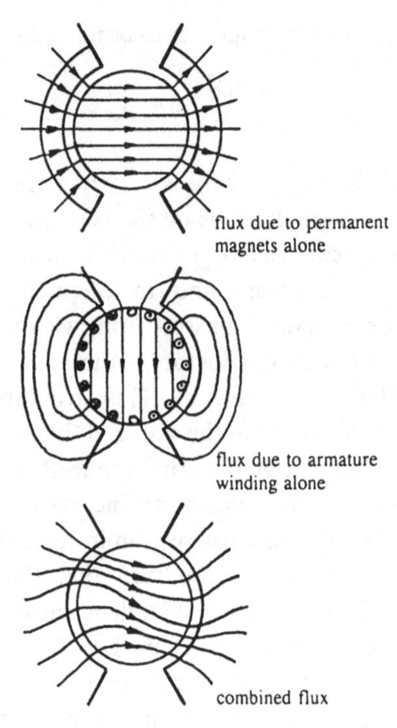

Figure 7.3. Components of the magnetic flux within a d.c. motor.

Every part of a magnet may operate under a different condition, but as was proved in Equation (4.25), provided that each part remains somewhere on the linear region of the demagnetization curve, then dynamic operation will not cause irreversible loss of flux. Given their characteristics, alnico magnets are much more susceptible to armature reaction than are higher coercivity types, such as ceramic ferrite or rare earth materials. However, although high coercivity in a magnet tends to indicate that it may better withstand irreversible changes due to armature reaction, one should caution that a short magnet length l_m will increase the excursions on the demagnetization curve, as shown in Figure 4.12. Because the armature reaction field is in "quadrature" with the main field, it is possible to adopt a motor design in which soft iron pole pieces are used to significantly isolate the magnets from the demagnetizing effect. Figure 7.4 shows some comparative geometries for a two-pole stator using different magnet types, pole pieces being employed in the case of the alnico design. Because soft iron acts as an equi-potential surface, pole pieces are also employed to ensure that magnet flux is distributed more uniformly around the air gap, to avoid possible localized saturation.

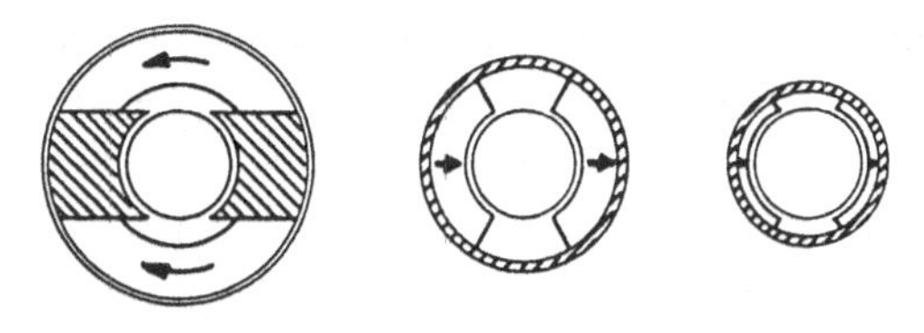

Figure 7.4. Permanent magnet stator layouts for a d.c. motor.

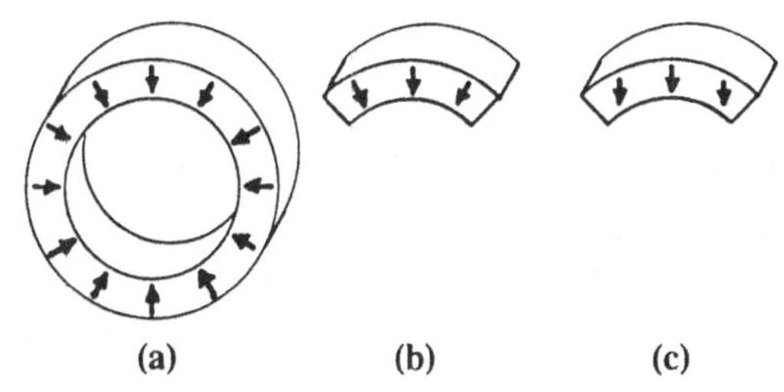

Figure 7.5. (*a*) A cylindrical magnet with radial orientation, and arc-shaped segments with (*b*) radial orientation and (*c*) uni-directional orientation.

The highest energy densities are obtained in anisotropic materials, whose orientation is imparted in the production process. The foregoing d.c. motor topologies show that cylindrical or arc-shaped magnets are frequently employed, whose preferred orientation is *radial*, facing the air gap. In Figure 2.22, we described the layout of an injection mold for producing radially oriented bonded magnet rings. This is much harder to achieve with fully dense sintered magnets, both because the required aligning field in the die is much greater, and because it must be uniform and radial across the cavity to produce a component such as Figure 7.5(*a*). One promising approach is the die-upsetting technique described in Section 2.6 that is used to manufacture rapidly quenched $Nd_2Fe_{14}B$ magnets; radial anisotropy is achieved by hot plastic deformation without an applied field, and the complete cylindrical shape is produced by extrusion. Even then, it is still necessary to subsequently magnetize the cylinder with a heteropolar array for use in a d.c. motor. As explained in Section 6.2, the width of these poles will be limited by the magnetizing fixture design, and the magnetic properties of the permanent magnet material. Consequently, for most fully dense materials, true radial orientation can only be achieved in arc-shaped segments such as Figure 7.5(*b*). Even then, the fixture providing the aligning field is frequently simplified to produce uni-directional orientation through the segment, as in Figure 7.5(*c*).

Figure 7.4 shows stators that are commonly used in d.c. commutator motors, as in Figure 7.1(*a*). The inverted topology of Figure 7.1(*b*) is most common for brushless d.c. motors, and comparative geometries using ceramic ferrite and sintered rare earth magnets are shown in Figure 7.6. In both commutator and brushless d.c. motors, there is a significant gain in magnet energy density by replacing ceramic ferrite with a sintered rare earth, which may be used to improve performance via a greater air gap flux density B_g. Even though this may lessen the electric loading in the armature winding, these comparisons show that the need to transfer flux between adjacent poles through the soft iron components permits only a slight reduction in rotor diameter D. However, this restriction does not apply to the rotor's length L, which may be reduced by around 45% (Howe, Birch and Gray, 1987).

We have been describing the most common layout for d.c. motors, known as the *cylindrical rotor*, in which the main field in the air gap passes substantially radially between the rotor and the stator. An alternative layout is widely used in *disk armature* motors, in which the main field passes *axially* between the planar surfaces of the rotor and stator, both of which are disk-shaped. A comparison between the two topologies is given in Figure 7.7, which shows cylindrical rotor and disk armature types of d.c. motor designed for the same application, in an automobile's radiator cooling fan. The permanent magnet field structure may comprise individual poles to one side of the armature, or duplicate sets of poles either side of the armature as shown in Figure 7.8. This topology generally embodies an "air gap" winding, one that resides in the air gap of the machine, not embedded in slots. The winding is therefore supported by a non-magnetic material such as epoxy resin, the free-standing armature in Figure 7.8 being encapsulated together with the commutator.

In a simpler construction, which has become popular for applications such as cassette tape drives, the winding is mounted directly onto a slotless

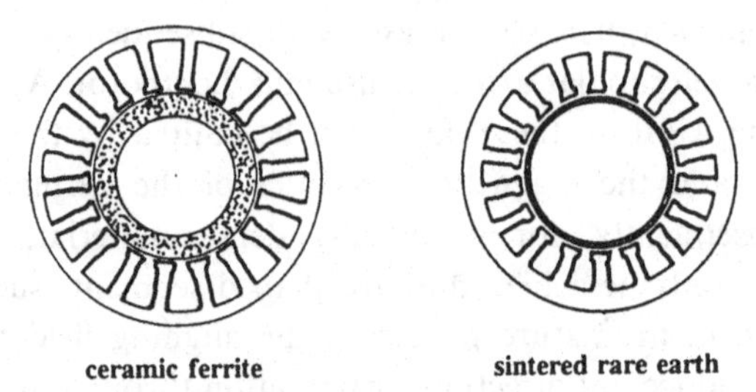

Figure 7.6. Permanent magnet rotor layouts for a brushless d.c. motor.

Figure 7.7. Automobile radiator cooling fan using a disk armature d.c. motor (left) and a cylindrical rotor d.c. motor (right).

soft iron support plate (Campbell, 1990). The general layout of this disk armature machine is shown in Figure 7.9, this being a brushless d.c. motor with Hall effect sensors to time the switching of the winding current. As with the cylindrical rotor machine, expressions can be written for B_g and A_c, now in terms of the inner and outer diameters D_1 and D_2 of the magnets, which define the air gap boundary (Campbell, 1974). In a similar manner to Equation (7.10), torque in a disk armature d.c. motor is found to depend on the same trade-off between the design of the magnetic circuit and the armature winding:

$$T=\left(\frac{\pi(D_2^2-D_1^2)D_1}{8}\right)B_gA_c \tag{7.11}$$

However, the absence of armature teeth and their incumbent saturation limit allows a higher air gap flux density to be employed, with consequent benefit to motor performance. Furthermore, when armature teeth pass under magnet poles, the changes in magnetic circuit reluctance cause

Figure 7.8. Exploded view of a disk armature commutator d.c. motor.

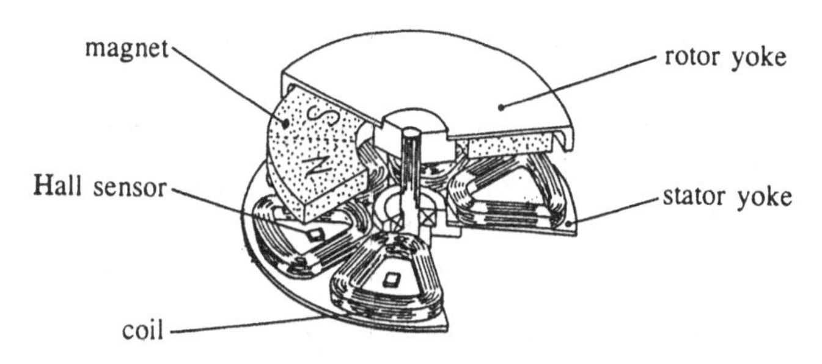

Figure 7.9. Layout of a disk armature brushless d.c. motor.

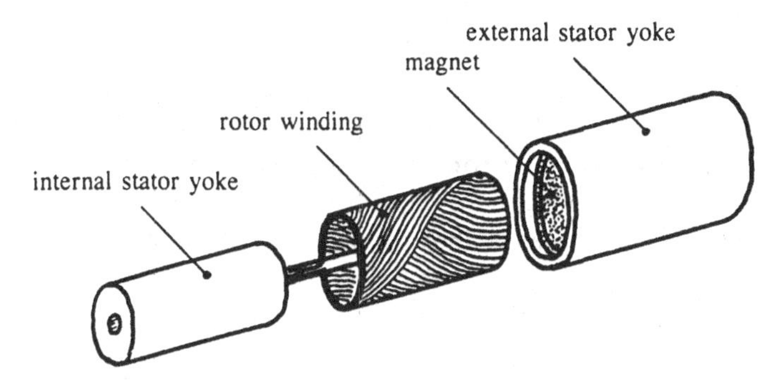

Figure 7.10. Layout of a slotless cylindrical rotor d.c. motor.

torque pulsations; this unwanted *reluctance torque* effect is absent in the disk armature topology. Figure 7.9 illustrates the use of a *single* toroidal magnet, onto which an array of poles is magnetized using a fixture such as Figure 6.9(*b*). This economy in assembly of a disk armature motor is clearly achieved with less restrictions than it was with a cylindrical magnet. However, the need to accommodate an "air gap" winding means there is a much longer air gap path in this magnetic circuit, and a greater magnet m.m.f. must be provided in a disk armature motor by using high coercivity permanent magnets – ceramic ferrite and rare earth types are generally employed. One benefit of the long air gap is that the winding's armature reaction field is ineffective.

A rotor comprising an encapsulated winding such as that in Figure 7.8 has been adapted to the cylindrical rotor topology shown in Figure 7.10 for both commutator and brushless d.c. motors. The cup-shaped rotor fits between interior and exterior stator components, either of which may contain the radially oriented cylindrical permanent magnet. Commonly known as a *slotless* d.c. motor, this layout is restricted to relatively lower power applications due to limitations in supporting the armature winding from one end alone. However, it shares some important advantages of the disk armature topology, including higher performance from an increased flux density without tooth saturation, and smooth rotation without reluctance torque pulsations. It also offers a low rotor inertia, lower than that of a comparable disk-shaped armature.

7.3 Stepper motors

In contrast to the d.c. motors described above, a stepper motor rotates in a sequence of discrete steps, each one initiated by exciting one phase of the stator armature winding, drawing the rotor into magnetic alignment with that phase. The alignment is enhanced by a lower magnetic reluctance in this condition, and so a variable reluctance in the magnetic circuit between the stator and the rotor is essential to the principle of operation of a stepper. The armature winding comprises multiple phases, and by exciting these in sequence, the rotor moves into alignment with each phase in a corresponding sequence, thus rotating in a series of steps. Rotation is characterized by torque pulsations, which correspond to the steps. Stepper motors may have any number of phases, but the simplest and most economical designs employ just two.

Operating in this manner, stepper motors are particularly amenable to digital control, which is usually performed open loop. However, closed loop control can also be employed in the event of dynamic stability problems, or to enhance the performance of a stepper motor system to the point where it can compete favorably with a commutator or brushless d.c. motor (Leenhouts, 1987). Steppers are mostly used in incremental motion applications, and are designed to increment their angular positions by a variety of step angles, limited only by the ability to construct teeth with an adequate variable reluctance. A common high resolution stepper achieves 200 steps per revolution, equivalent to a step angle of 1.8°. The major markets utilizing this motor characteristic include computer peripherals, plotters, industrial drives and consumer products. In printers, steppers are used for paper and ribbon advance, print head positioning, and daisy-wheel rotation. They are commonly used as head actuators in both floppy and hard disk drives, competing with voice coil actuators for the latter. Machine tools and robots use higher torque stepper motors for a variety of positioning requirements. Miniature stepper motors are used extensively in electronic cameras, especially for lens focussing.

Stepper motors do not necessarily employ permanent magnets, because the required variable reluctance can be established between a toothed stator structure carrying the armature winding and an entirely passive toothed soft iron rotor. In a *variable reluctance* stepper, the armature winding is solely responsible for producing the air gap flux, but a more efficient design is achieved by using a permanent magnet to assist in this flux generation. The torque generated by a stepper motor may be evaluated from the air gap energy using

$$T = -\frac{dE_g}{d\theta} \tag{7.12}$$

As discussed in Section 4.5, there are various ways to express E_g, one that includes contributions from a permanent magnet and a coil being

$$E_g = \frac{-V_m B_m H_m}{2k_1 k_2} + \frac{\lambda i}{2k_1 k_2} \tag{4.36}$$

Clearly the air gap energy, and hence the torque, is raised by increasing the contribution from the armature winding (via i), or from the permanent magnet material (B_m, H_m), which provides the steady bias field. Once again we find that there is a trade-off between the design of the permanent magnet circuit and the armature winding.

In a *hybrid* stepper motor, permanent magnets are incorporated into the rotor, sandwiched between toothed soft iron rotor cups. Multiple sections are frequently used to increase the torque, as shown by the rotor in Figure 7.11. Each magnet has a very simple disk shape, uniformly magnetized with a single pole through its thickness. To illustrate the principal of operation, consider that one of the rotor cups in Figure 7.11 is established as a *North* pole by the permanent magnets, and this aligns with a two-phase toothed stator structure as shown in Figure 7.12. Only one of the winding phases is energized, the direction of the coil currents being shown on alternate stator poles. The fields due to the permanent magnet and the armature winding reinforce each other in the gaps under poles 1 and 5, whereas they oppose each other under poles 3 and 7. Furthermore, the stator structure is designed in such a way that its teeth align with the rotor teeth under poles 1 and 5, but are misaligned under poles 3 and 7. These effects combine to establish much stronger fields in the gaps of poles 1 and 5, the higher energy of which creates a net torque that holds the rotor in this position. This *holding torque* is a most fundamental characteristic of a stepper motor. Because the stator teeth extend over the full length of the rotor, the alternate rotor cups with *South* poles must have their teeth offset by half a tooth pitch, which in the same angular position are aligned under stator poles 3 and 7. Rotation is

Figure 7.11. Rotor of a hybrid stepper motor. (Courtesy of The Superior Electric Company.)

Figure 7.12. Rotor alignment with one phase in a hybrid stepper motor.

initiated by energizing the other winding phase with a current direction that causes a torque to now align the rotor teeth under stator poles 2 and 6 – there is a net rotation of the rotor by a quarter of a tooth pitch. Next, the original phase is energized with current in the opposite direction to that shown in Figure 7.12, so the rotor teeth align under poles 3 and 7, and so on.

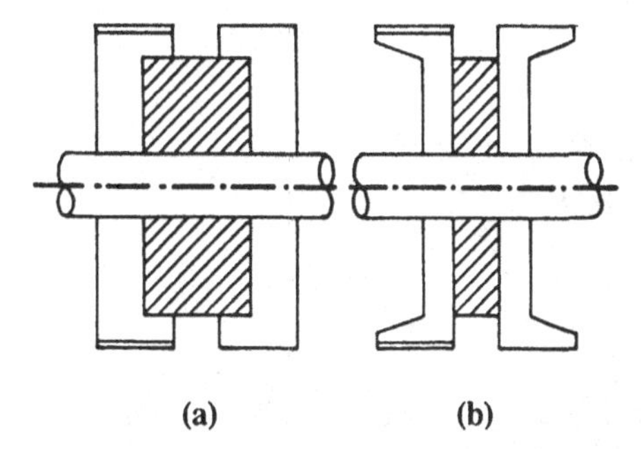

Figure 7.13. Cross-sections of hybrid stepper rotors with (*a*) Alnico 5 and (*b*) sintered $Nd_2Fe_{14}B$.

Alnico magnets are commonly used in hybrid steppers, although some alternative designs now use Sm–Co or Nd–Fe–B. Figure 7.13 shows two alternative rotors having a single magnet between two cups, in which a clear advantage of sintered $Nd_2Fe_{14}B$ over Alnico 5 is that it provides a much reduced inertia. In the magnetic circuit of a hybrid stepper, the magnet and the winding are in *series*. When exciting the armature winding either to reinforce or to oppose the permanent magnet's field in the air gap, it is most undesirable to consume electrical power to initiate dynamic operation by changing the magnet's operating point, but if this is unavoidable, then it is beneficial to be using a material such as alnico with a recoil permeability $\mu_{rec} \gg 1$ (for example, $\mu_{rec} \approx 4$ in Alnico 5). With a high coercivity rare earth magnet, the stator and rotor structures must be designed so that the magnet's operating condition does not change – for the situation in Figure 7.12, the decrease in flux under poles 3 and 7 should equal the increase under poles 1 and 5. The armature winding is being used to *switch* the magnet's field to different angular positions around the air gap. The design objective is to maximize this difference in the air gap field for the greatest holding torque, without significantly changing the magnet's condition.

A different type of permanent magnet stepper motor, which has a very simple assembly and a much lower manufacturing cost, is known as the "*can-stack*" motor. The toothed soft iron cups are no longer present in the rotor, which is now a simple cylindrical permanent magnet with a shaft passing through it. This is usually made from an *isotropic* grade of ceramic ferrite, Sm_2Co_{17} or $Nd_2Fe_{14}B$, to facilitate a heteropolar structure being magnetized directly into its surface as illustrated in Figure 7.14. Furthermore, this is normally also a compression- or injection-molded material, in keeping with applications requiring production in high volume

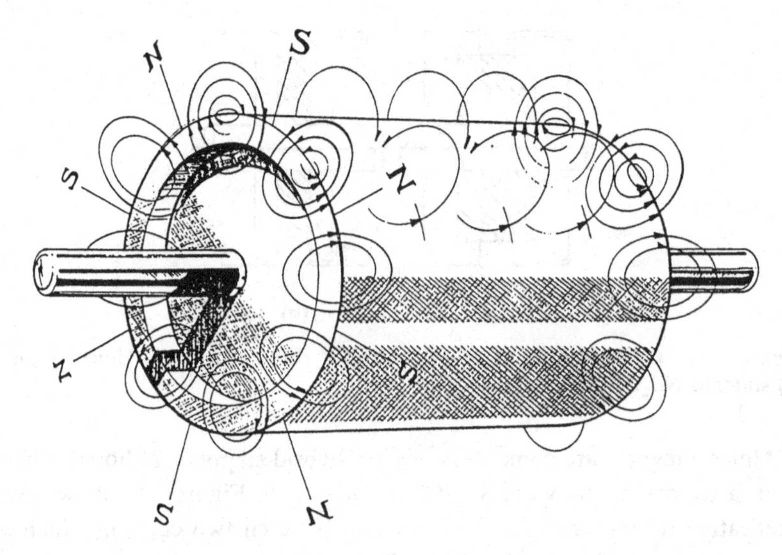

Figure 7.14. Isotropic permanent magnet rotor, and flux lines showing heteropolar structure.

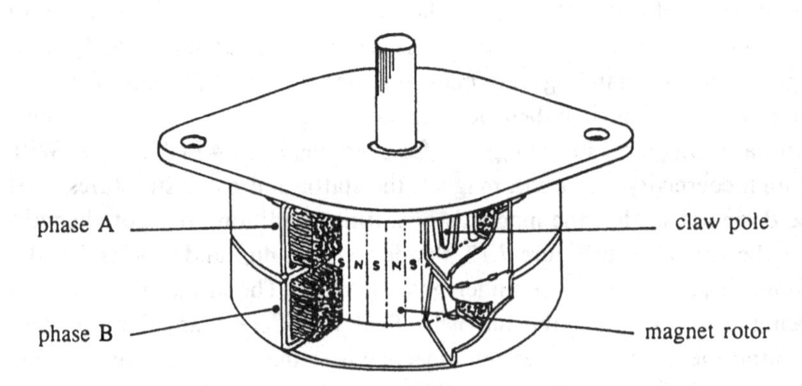

Figure 7.15. Layout of a permanent magnet, two-phase claw pole *can-stack* stepper motor.

at low cost. A photograph of such an injection-molded ferrite magnet rotor was shown in Figure 2.23.

The can-stack motor has a two-phase winding, each phase being a simple bobbin-wound coil. Each coil is surrounded by a soft iron core which has a "claw-pole" structure in its bore, facing the rotor across the air gap. As shown in Figure 7.15, the two winding phases are separate assemblies, stacked axially beside each other over the rotor, but with the teeth in the phases offset by half a tooth pitch. Energizing one phase sets up an alternating *North–South* pole array on its claws, and a holding torque is created to align the permanent magnet poles on the rotor. Rotation is initiated by energizing the other phase, creating a torque which turns the rotor half a tooth (or pole) pitch. The sequence continues by next exciting the original phase in the opposite direction, and so on. The torque contributions from the two phases are quite independent, and it is therefore possible to excite both phases together to introduce intermediate stable positions with a step angle of a quarter of a tooth pitch. As illustrated in Figure 7.14, the magnetization lines up along approximately circular paths between adjacent poles within the permanent magnet rotor; the higher the number of poles on a given rotor, the shorter will be the equivalent magnet length, leading to a correspondingly smaller m.m.f. and air gap flux density. Therefore, while increasing the number of poles raises the resolution of a can-stack motor, this also reduces its holding torque.

In the previous section, we described an alternative *disk-armature* layout for a d.c. motor, in which the main field passes axially between the planar surfaces of the rotor and stator. A similar transformation has been made to the permanent magnet stepper motor – instead of a cylindrical magnet, the rotor is a thin disk of axially oriented anisotropic material, usually fully dense Sm_2Co_{17} or $Nd_2Fe_{14}B$. The principle of operation is illustrated in Figure 7.16, which shows only *one* of the C-shaped stator cores belonging to each of the two phases – all the cores in one phase have a similar angular orientation with respect to the magnetized poles in the permanent magnet rotor. Exciting the phases alternately causes the rotor

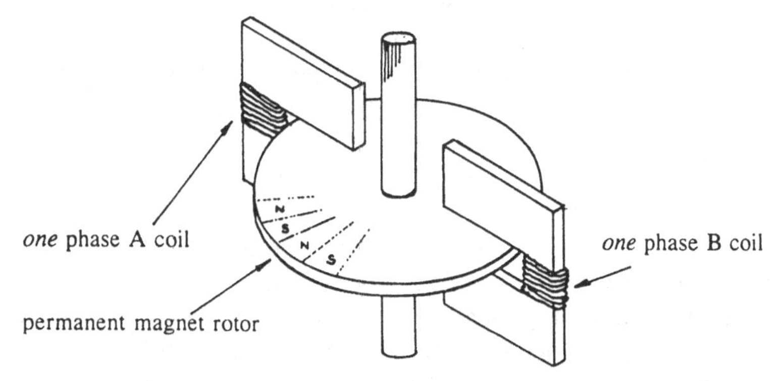

Figure 7.16. Principle of operation of a permanent magnet disk-rotor stepper motor.

to perform a stepped rotation, in the same way as described for the can-stack motor.

As before, the resolution of this motor is determined by the density of poles that can be magnetized on the rotor. It is the axial orientation of the disk-rotor that allows a high coercivity anisotropic magnet to be utilized, which can be more easily magnetized with a high density array. High coercivity also provides a much greater m.m.f. than the isotropic grades used in the can-stack motor, giving a higher air gap flux density and holding torque. Furthermore, the inertia of the disk-rotor is very low – compare that in Figure 7.16 with the hybrid permanent magnet rotor in Figure 7.11. High torque and low inertia combine to give this special type of stepper motor the advantage of excellent acceleration, an important characteristic in many servo control applications.

7.4 Synchronous motors

In a d.c. motor, the flux produced by the armature winding is in *quadrature* with the main flux from the permanent magnets, as illustrated in Figure 7.3. This 90° phase relationship is maintained by mechanical or electronic commutation of the winding. The layout of a brushless d.c. motor in Figure 7.1(*b*) may equally function as an a.c. *synchronous* motor, if the multi-phase stator winding produces an armature field, which rotates around the air gap. No commutation is needed, and the permanent magnet rotor will rotate in synchronism with the armature field, producing maximum torque when the two components of flux are in quadrature. In fact, the characteristic power equation for a synchronous machine (per phase) is expressed as

$$P = \frac{Ve}{X} \sin \delta \tag{7.13}$$

The inductance of each winding phase is now included in its reactance X. δ is the *torque angle* between the flux vectors, confirming maximum power at $\delta = 90°$.

Synchronous motors with permanent magnet field systems offer performance advantages over other a.c. machines both for fixed speed applications operating at a constant supply voltage and frequency, and for variable speed drives where control is provided by a variable frequency inverter. These advantages include higher efficiency and power factor, and a greater power density in the machine. However, while a permanent magnet field system provides the best *synchronous* operating performance, the starting torque is somewhat weak – just consider that δ is never

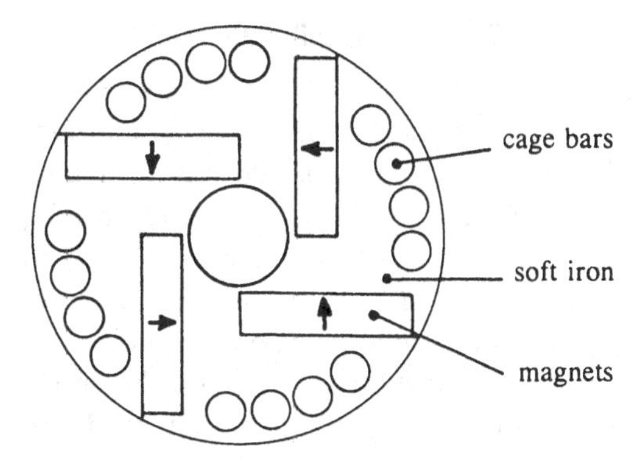

Figure 7.17. A typical four-pole rotor of a line-start synchronous motor.

constant in Equation (7.13) during *a*syncronous running. The best starting torque is achieved in an a.c. induction motor, and an attempt to combine the most attractive features of each machine gave rise to the *line-start* synchronous motor. Figure 7.17 shows a typical rotor, including both permanent magnets and squirrel-cage induction bars, from which it is clear that the line-start motor has a relatively complex construction.

During starting and the initial run-up to speed, the rotating stator winding flux induces currents in the short-circuited induction bars of the squirrel-cage, which react with the stator field to accelerate the rotor towards synchronism. Once the speed of the rotor achieves synchronism with the stator field, there is no longer relative motion between the two, and therefore no induced currents. The flux from the permanent magnets now takes over, and the machine functions as a conventional synchronous motor. Approaching synchronous speed, however, the permanent magnets embedded in the rotor can experience a large demagnetizing effect from the armature winding flux, so only high coercivity ceramic ferrite and rare earth types with linear demagnetization characteristics are employed.

One type of permanent magnet synchronous motor that does not require assistance with its starting torque is the *hysteresis* motor. The layout of Figure 7.1(*b*) is once again applicable, except that the rotor magnet is a *low* coercivity material whose magnetization state can be influenced by the armature winding field – this magnetizes the rotor. The hysteresis in its B *versus* H loop is used to create a phase difference between the armature winding and rotor magnet fluxes, which will now always rotate in synchronism with the rotor lagging by a torque angle δ. This is a simple motor, which is widely used for fixed frequency timing applications such as clocks.

7.5 Moving coil actuators

Moving coil actuators were originally used only as the voice coils in loudspeakers and similar devices. They have now become increasingly popular for a wide range of actuation needs, but are frequently still called *voice coil* actuators. Typical applications include head positioning in computer disk and tape drives, servo controlled valves, and mirror positioning in laser scanners. Moving coil actuators compete favorably with d.c. and stepper motors in applications requiring precise control of position over a relatively short stroke, because they do not exhibit the backlash, irregular motion or power loss that result from converting rotary to linear motion.

Figure 7.18 shows two possible layouts for a moving coil actuator, one in which the magnet is a simple disk, the other using a radially oriented magnet. The soft iron pole structure including the permanent magnet is fixed; the coil, whose suspension is not shown in Figure 7.18, moves linearly along the air gap. The coil contains no iron, so its mechanical rigidity is provided either by encapsulation in epoxy resin, or by winding it on a thin, non-magnetic bobbin. Consequently, the coil's motion is smooth because there are no variable magnetic reluctance effects to cause force pulsations, and there are no hysteresis power losses. The moving coil has a very low mass, giving the actuator a rapid response time. This is further enhanced by the *air gap* coil having a low inductance, allowing a high rate of change of current to be applied to it. Magnetic flux must be sustained over a relatively long air gap, however, which favors use of high coercivity magnets. As is shown later, the performance of a moving coil actuator is enhanced by increasing the air gap flux density – one way to achieve this is to focus flux from a large magnet into a small gap, as shown in Figure 7.18(*a*).

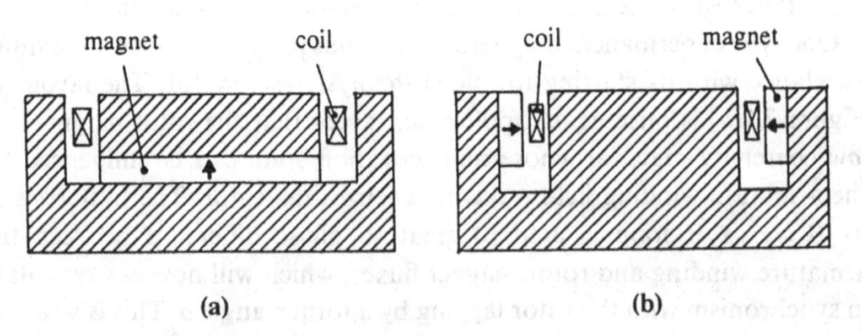

Figure 7.18. Alternative layouts of a moving coil actuator, using (*a*) a disk-shaped magnet, and (*b*) a radially oriented cylinder.

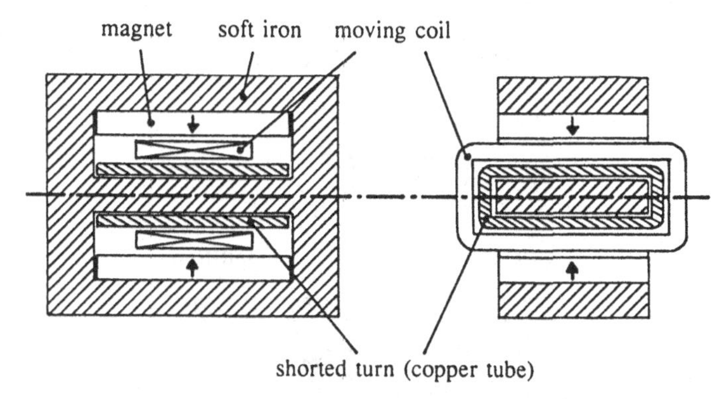

Figure 7.19. Layout of a moving coil linear actuator.

Motion of the moving coil actuator need not be strictly linear, but may be constrained over a limited arc of rotation – this is common in disk drive head actuators, where the coil only rotates a few degrees, but moves the head over the width of the platters through a mechanical lever arm. The magnetic circuit of this voice coil actuator is shown in Figure 7.19. Two high coercivity permanent magnets direct flux into the central limb of a soft iron core, and the flux returns to the magnets via two end plates and the outer soft iron limbs. The coil is mounted to the mechanical lever arm (not shown in Figure 7.19), which supports its travel along the central limb under the magnets. This is exactly the example whose flux distribution was analyzed in Figures 5.6–5.9 using the finite element method. A practical difficulty with this topology is that the coil itself experiences a *closed* magnetic path through the soft iron core (which does not include the permanent magnets), as is apparent from the flux plot of Figure 5.8. In this case, the coil's inductance is no longer very low, and the longer electrical time constant inhibits the actuator's ability to achieve high performance. A technique that is commonly employed to reduce the inductance is to add a shorted turn in the form of a copper tube over the central limb, closely coupled to the moving coil as shown in Figure 7.19. This creates a transformer effect within the coil, and reduces its inductance typically by a factor of 40 (Wagner, 1982).

An alternative topology for the moving coil actuator involves the flat coil shown in Figure 7.20, which moves in a limited arc with each side under two adjacent magnets. This layout is simply a sector of the disk-armature d.c. motor that was depicted in Figure 7.9. This coil does have an inherently low inductance, because it experiences a large air gap comprising the main gap and the magnets themselves. However, the

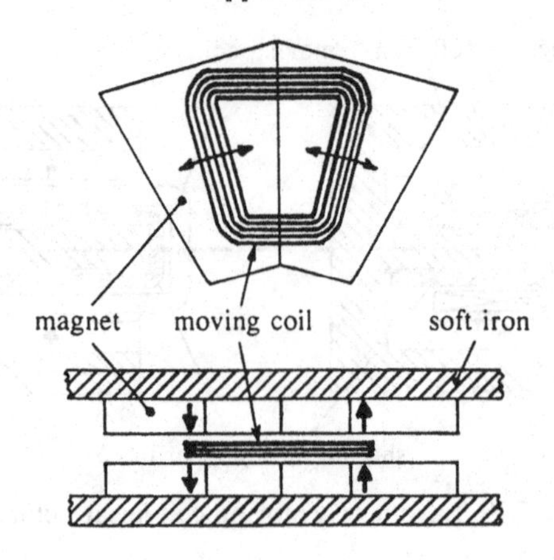

Figure 7.20. Layout of a flat voice coil actuator.

previous layout of Figure 7.19 is usually preferred for dynamic applications, even at the cost of including a shorted turn, because it provides greater structural rigidity for the coil.

Moving coil actuators are frequently employed for their ability to achieve high performance, such as in disk drive head positioning where minimum time to "access" data on the platters is desired. Dynamic motion of the coil in an arc without damping is expressed in terms of the torque T, angular acceleration α, and moment of inertia J:

$$T = J\alpha \tag{7.14}$$

Usually, rotation through an angle θ (radians) is required to be completed within a time t, for which

$$\theta = \frac{\alpha t^2}{2} \tag{7.15}$$

Combining Equations (7.14) and (7.15),

$$\theta = \frac{Tt^2}{2J} \tag{7.16}$$

The similarity between a moving coil actuator and a d.c. motor has already been noted, and the same definition for the torque constant K_T is used for each device. Thus, using Equation (7.7) in (7.16),

$$\theta = \frac{K_T i t^2}{2J} \tag{7.17}$$

Rotation of the coil through a limited arc requires both to accelerate and to decelerate, for which similar equations can be written. To simplify understanding of its performance, consider that the coil accelerates and decelerates in equal time periods with the torque, angular velocity and displacement profiles shown in Figure 7.21. Equation (7.17) then applies to each of these periods, and the total rotation θ_{m} is related to the total time to move t_{m} via

$$\frac{\theta_{\mathrm{m}}}{2}=\frac{K_T i}{2J}\left(\frac{t_{\mathrm{m}}}{2}\right)^2 \tag{7.18}$$

Hence, the move time is expressed as

$$t_{\mathrm{m}}=2\left(\frac{\theta_{\mathrm{m}} J}{K_T i}\right)^{1/2} \tag{7.19}$$

The requirements for a moving coil actuator specify the rotation θ_{m}, and the mechanical support system largely determines the inertia J. To minimize the "access" time in a disk drive, for example, means minimizing t_{m} by making the $K_T i$ product as large as possible. K_T may be derived from the design parameters of the actuator, but since we have already derived similar relationships for a d.c. motor in Section 7.2, we shall consider the actuator as a motor with just *one* pole pair and *one* coil. Noting that $K_T=K_{\mathrm{E}}$ and that $a=1$, the torque constant for a moving

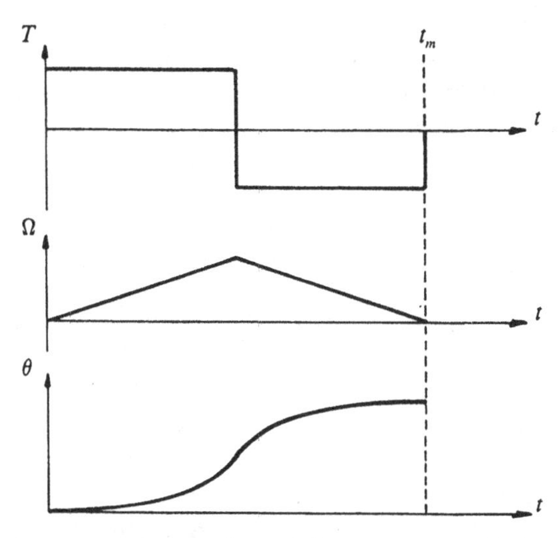

Figure 7.21. Torque, angular velocity and displacement profiles for a moving coil actuator.

coil actuator is found by combining Equations (7.2) and (7.8):

$$K_T = N(DL)B_g \tag{7.20}$$

The moving coil of N turns has conductors of active length L, which lie at an effective diameter D in this limited rotation device. To increase K_T by merely raising the number of turns will also raise the coil's resistance and inertia. The more common way to improve the torque constant is to increase the air gap flux density B_g – this is exactly as was concluded for the permanent magnet d.c. motor. Of course, the procedure for complete design of a voice coil actuator is somewhat more complex than this. For a value of K_T established by the magnetic circuit layout, the required current is then determined by Equation (7.19), and i will in turn specify the coil resistance R, consistent with any limitations on temperature rise in the device. Clearly R is a dependent parameter, which also affects the performance of the actuator through the electrical time constant and the coil inertia.

We conclude this section with a mention of *loudspeakers*, the application that spawned voice coil actuators. The design goal in a loudspeaker is to convert the electrical input power Vi most effectively into mechanical power Fv. Coil heating i^2R is subtracted from Vi to yield ei, the "electromagnetic" power available for conversion into force. In a loudspeaker, it is desirable to make R large, and the corresponding reduction in i leads to the most efficient conversion of energy. Since the voice coil is attached to the speaker cone, its force F translates directly into the sound pressure generated. Once again, F is directly proportional to B_g, and the dynamic performance of a loudspeaker depends critically upon the magnitude of air gap flux density that can be established. This also maximizes the magnetic field energy in the air gap via

$$E_g = \frac{V_g B_g^2}{2\mu_0} \tag{4.30}$$

Loudspeakers rarely use a radially oriented permanent magnet adjacent to the air gap as was shown in Figure 7.18(*b*) – they usually adopt the more economic approach of employing simple disk-shaped magnets with axial anisotropy. Pole pieces are used to focus the magnet flux to a high density in the radial air gap, as shown in the alternative layouts with different magnet materials in Figure 7.22. These three magnet structures produce approximately the same air gap energy, from which the benefit of using Sm_2Co_{17} or $Nd_2Fe_{14}B$ for miniaturization is apparent. Because of

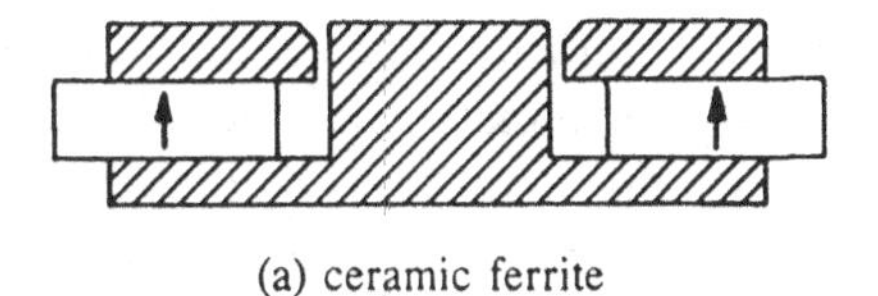

(a) ceramic ferrite

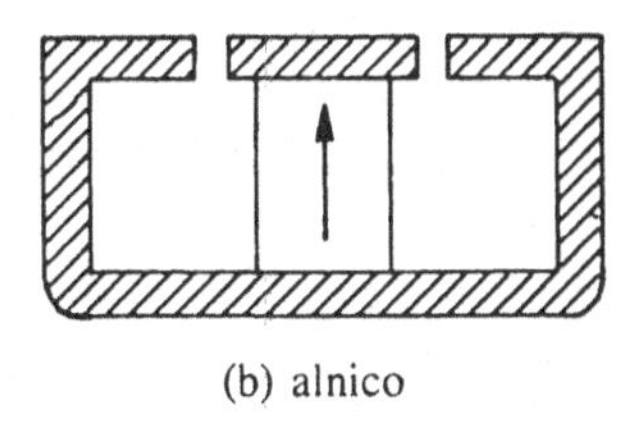

(b) alnico

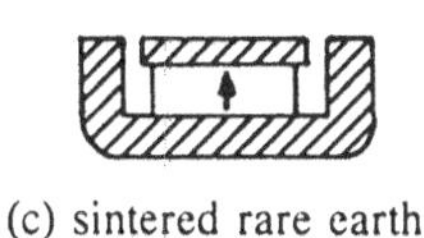

(c) sintered rare earth

Figure 7.22. Alternative magnet structures for a loudspeaker.

the low flux density in a ceramic ferrite, it is common for this to be an external cylindrical disk as shown, to utilize a greater magnet surface area.

7.6 Holding force actuators

In Section 7.5, one special class of magnetic actuator having a moving coil was described. There are many other types, which employ permanent magnets, that have other components to provide motion; this is usually some part of the soft iron core in the magnetic circuit. Consider, for example, that a flux density B_g is established in the air gap of area A_g between two pole faces in a magnetic circuit, as was shown in Figure 4.18. We derived the force of attraction between these poles as

$$F_x = -\frac{A_g B_g^2}{2\mu_0} = -\frac{\Phi^2}{2\mu_0 A_g} \qquad (4.47)$$

While the length of the gap does not appear in Equation (4.47), the *holding* force does depend on this via the magnetic circuit equations: the longer the gap becomes, the lower will be the flux density, and hence the force. This characteristic is particularly useful in linear actuators, which require only limited control over their stroke, and it has formed the basis for operation of the *impact* hammer actuators that are used in computer printers. In the line printer, hammers strike the paper and ribbon against a metal band containing the characters. In the dot matrix printer, an array of hammers can print a line of dots, and the actuators are selectively activated to form characters as the head assembly traverses the page.

A typical layout for the *holding* force actuator in a print hammer is shown in Figure 7.23. The field from a permanent magnet passes through a spring, then through an air gap and back into the soft iron core. In Figure 7.23, the spring is shown in its neutral (undeflected) position, but the force of attraction across the air gap is sufficient to deflect the spring against the pole piece, closing the gap. The holding force, calculated from Equation (4.47), will increase with deflection δ, closing the gap and raising the flux. This force must overcome the spring force, given in terms of the spring constant κ as

$$F_{\mathrm{k}} = \kappa\delta \tag{7.21}$$

To latch the spring, $|F_x|$ must exceed F_{k} by a comfortable amount, as shown in Figure 7.24. The magnetic circuit also contains a coil on the pole piece, which may be energized momentarily to oppose the field in the air gap due to the magnet. This reduces B_{g} such that $|F_x| < F_{\mathrm{k}}$, releasing the spring to print a dot (or character).

Exciting the coil does not actually change the operating point of the permanent magnet very much; rather, it diverts the magnet's flux away from the main air gap into leakage paths. The electrical power input to the actuator may therefore be quite small, both because of this and because the energy in the coil does not directly produce the actuation force. Contrast this to the moving coil actuator described in Section 7.5, in which the coil is directly involved in the energy conversion process. In

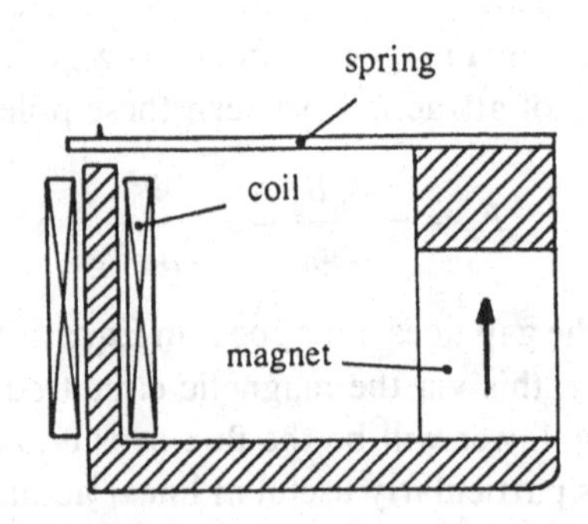

Figure 7.23. Layout of a print hammer actuator.

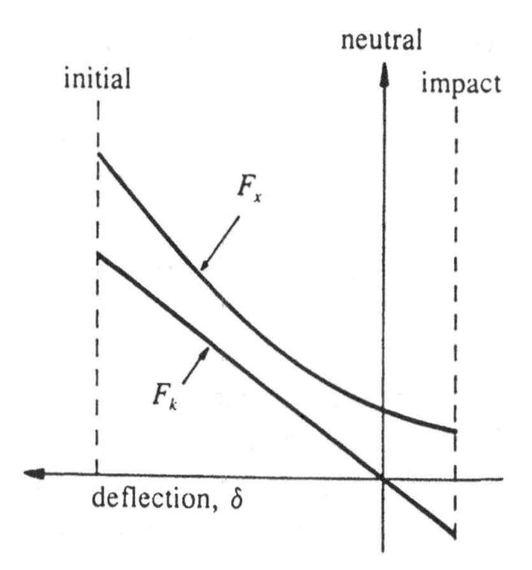

Figure 7.24. Force *versus* displacement profile for a hammer actuator.

this holding force actuator, the coil is only excited for a small portion of the cycle of operation, just long enough to release the spring. The spring then follows an equation of motion in x, its distance from the pole piece; if the neutral is at $x = x_0$, then $\delta = x_0 - x$, and the undamped motion is described by

$$-F_x + \kappa(x_0 - x) = m_e \frac{d^2x}{dt^2} \tag{7.22}$$

where m_e is the equivalent mass of the spring. This expression has a natural frequency ω_0, which is found from Equation (7.22) to be

$$\omega_0 = \left(\frac{\kappa}{m_e}\right)^{1/2} \tag{7.23}$$

With the assumption that there is no damping present, the spring's motion is oscillatory about its neutral position at frequency ω_0, except of course that this is interrupted by impact. Thereafter, because the coil's excitation has been removed, $|F_x| > F_k$ as shown in Figure 7.24, and the air gap flux is sufficient to re-latch the spring to the pole piece, in readiness for the next current impulse to the coil.

The frequency of operation of this actuator is not limited by the magnetic circuit design, because coil excitation does not drive the spring throughout its cycle; this limitation is imposed by the natural frequency of the spring as given by Equation (7.23). To increase ω_0, the spring should be made stiffer, but this in turn requires a greater magnetic latching force, and a corresponding increase in air gap flux according to Equation (4.47). The actuator layout shown in Figure 7.23 has an inherent disadvantage in this regard, because the spring itself is a component of the magnetic circuit, which must carry the magnet flux to the air gap. In higher performance actuators of this type, the spring is absolved of this dual role, by attaching it to an armature, which replaces it in the main flux path. More complex versions of this actuator are becoming common for controlling valves, an example shown in Figure 7.25 being for the control of fluid (Campbell, Tsals and Matsuura, 1986). This device uses a sintered Sm_2Co_{17} or $Nd_2Fe_{14}B$ magnet, and includes a very well-defined leakage gap into which the magnet flux is diverted when the coil is excited; the objective is to release the armature (and open the valve) with a low electrical power requirement, so the magnet's operating point should not change significantly throughout the cycle of operation.

An analysis of a simple permanent magnet circuit of Figure 4.20 was performed as an example in Section 4.6, and since there is no coil present, the magnet's operating point does change due to armature movement alone in the manner shown in Figure 4.21. This layout forms the basis of holding force actuators, which are used for workholding applications, such as magnetic chucks. The workpiece that is to be held must be of soft magnetic material, because it becomes the *armature* of the actuator. In some magnetic chucks, the field source is a coil rather than a permanent magnet, but while this facilitates control of the holding force, it is not fail-safe – loss of electrical power will release the workpiece. The field from a permanent magnet must be diverted from the armature to release it, but rather than using a coil to do this as in Figure 7.25, it is more usual to physically change the stator assembly in a magnetic chuck.

There are many innovative designs of workholding devices using permanent magnets, but the example of a flat magnetic chuck in Figure

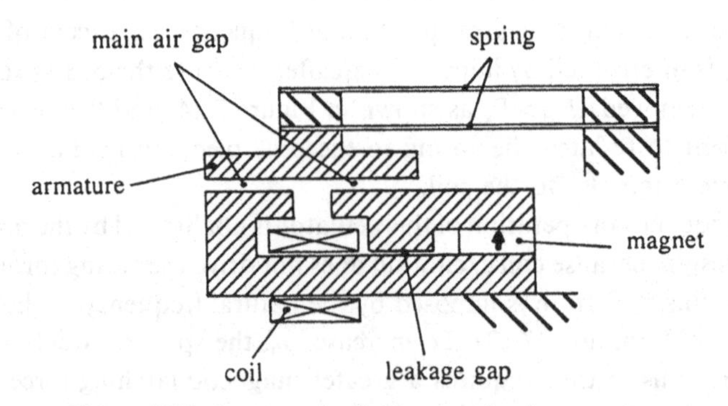

Figure 7.25. Layout of a fluid control valve actuator.

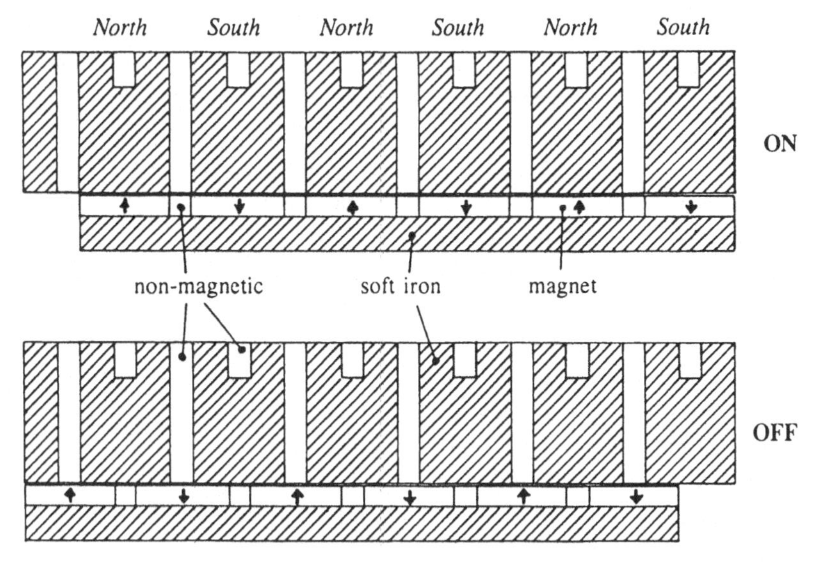

Figure 7.26. Layout of a flat magnetic chuck.

7.26 serves to illustrate how the field is diverted mechanically from a soft iron workpiece. Permanent magnets are mounted onto a backing plate, which slides under the pole piece assembly in the stator. In the ON position, free *North* and *South* poles exist at the upper surface of the stator, to which the workpiece will be held. In the OFF position, the magnet assembly is moved so that the field is mostly contained by short-circuit paths within the stator. According to Equation (4.47), high air gap flux density is the key to achieving a high holding force, so the use of ceramic ferrite magnets is frequently accompanied by some arrangement for flux focussing in the stator assembly. With inherent operation under dynamic recoil, the linear demagnetization characteristic is an important advantage of sintered rare earth magnets in workholding devices, in addition to their high energy densities. High remanence alnicos also appear well suited to this application, but a careful analysis of their dynamic operation must be performed in the manner of Figure 4.21.

7.7 Magnetic suspensions

If the armature in the magnetic circuit of Figure 4.20 is part of a much longer rotating shaft, then the permanent magnet stator will exert a force pulling the shaft off its axis of rotation. Add a similar stator, which is diametrically opposed to the first, and the shaft will be in a state of unstable equilibrium when the air gaps on either side are equal. This layout forms the basis of *magnetic bearings*, which have become an important alternative to conventional ball or roller bearings for ultra-high speed drives. For example, the rotor of a brushless d.c. motor shown in Figure 7.1(*b*) is entirely passive, and its upper speed limit can be raised substantially if physical contact in the bearings is eliminated. Whether the air gap flux creates forces of attraction between the components, as described in the Section 7.6, or forces in repulsion, as discussed below, a fundamental principal known as *Earnshaw's Theorem* states that stable equilibrium can only be achieved if active control is maintained over at least one axis (Earnshaw, 1839). This means replacing the permanent magnets with electromagnetic coils, whose excitations are servo-controlled using sensors to measure misalignment of the shaft. In practice, it is more economical and efficient to combine permanent magnets and coils, such that the magnets' fields provide the basic forces supporting the shaft. With these changes made to the circuit of Figure 4.20, a simple magnetic bearing layout is shown schematically in Figure 7.27.

The radial bearing in Figure 7.27 is but one of a number of possible layouts, derived from the orientation of concentric permanent magnet rings. Just a few of these layouts are illustrated in cross-section in Figure 7.28, the rings in (*a*) and (*b*) being repelled from each other, those in (*c*)–(*f*) being attracted; (*d*) is the configuration used in Figure 7.27, but with a magnetic steel shaft substituted for the inner magnet. A magnetic bearing may equally well be used to control the position of a shaft along its axis of rotation. Figure 7.29 shows some alternative configurations for

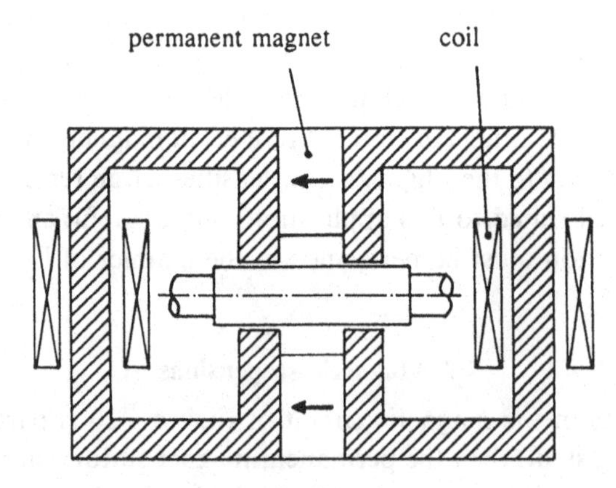

Figure 7.27. Layout of a radial magnetic bearing with permanent magnet bias fields.

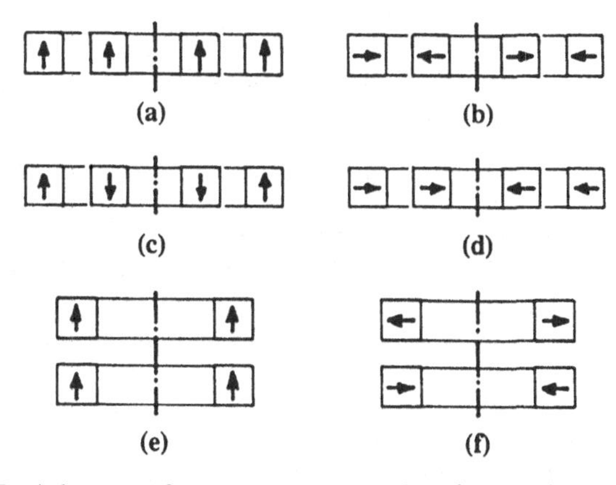

Figure 7.28. Basic layouts of permanent magnet rings for a radial magnetic bearing.

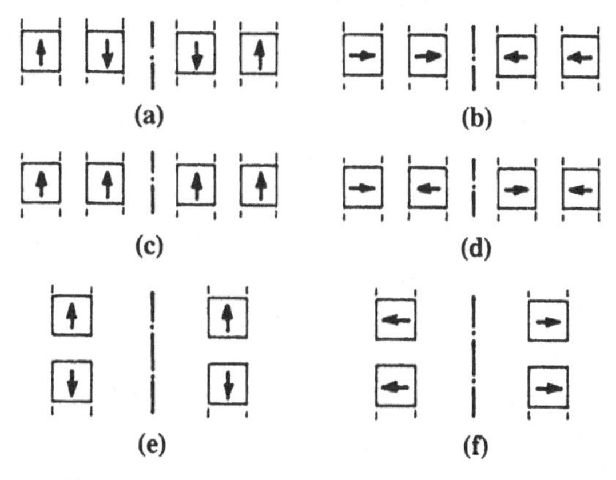

Figure 7.29. Basic layouts of permanent magnet rings for an axial magnetic bearing.

axial bearings, which are derived by reversing the magnetization in one of the rings in each layout of Figure 7.28.

In those magnetic bearings that work in repulsion mode, like layouts (*a*) and (*b*) in Figure 7.28 and (*c*)–(*f*) in Figure 7.29, the opposing permanent magnets subject each other to quite large demagnetizing fields, which can only be withstood in high coercivity materials having demagnetization characteristics that are substantially linear. While ceramic ferrites are sometimes used, rare earth magnets are ideal in this regard, their high energies also providing large support forces. A particular difficulty with magnets in repulsion is that the air gaps are quite large, and it is difficult to assign unique dimensions to the flux paths that would allow the use of magnetic permeances. The only solution is to calculate the flux distribution using magnetic potential in an equation such as

$$\nabla \times \frac{1}{\mu_0 \mu} \nabla \times \boldsymbol{A} = \boldsymbol{J} \tag{5.28}$$

where the magnetization sources are included in $\boldsymbol{J}$. This may be solved by the finite difference or finite element methods, the flux density then being calculated from the potential distribution, and the forces determined. Alternatively, it is possible to use the integral form of a differential field equation such as (5.28), summing the effects of the magnetization vectors (Yonnet, 1981).

Magnetic bearings can also be used to support linear systems, the most notable example being in *magnetic levitation*, or "*maglev*". Maglev has been implemented for suspension of both low and high speed ground vehicles, from urban transportation to rocket sleds. In most common forms of maglev, the field source is carried in the moving vehicle, because it would be impractical to distribute this along the full length of track. Early maglev systems employed low temperature superconducting coils, whose high fields provided the required lift forces for realistic full-scale vehicles. Later, developments in high temperature superconductors alleviated some of the difficulty in running cryogenic hardware aboard a moving vehicle, but the evolution of high energy rare earth permanent magnets also made them eminently suited to this application.

Urban transport vehicles that run at speeds up to around 120 km h^{-1} typically use a maglev system with sintered Sm_2Co_{17} or $Nd_2Fe_{14}B$ magnets in the vehicle, providing a lift force through their attraction to a long steel plate in the track. Each side of the vehicle is supported as shown in Figure 7.30. As was discussed for the magnetic bearing in Figure 7.27, it is necessary to supplement the permanent magnets' force with active control, to stabilize and maintain the air gap clearance. In this case, adjustment of the gap is provided by spacer wheels in a servo-controlled mechanical suspension, located at each corner of the vehicle (not shown in Figure 7.30). The control is designed so that the maglev suspension carries 90–95% of the vehicle load, leaving only 5–10% through the wheels, which benefits the efficiency and reliability of the mechanical system. Figure 7.30 also shows that the steel track plate has a slotted surface, which carries an armature winding. This operates as a three-phase a.c. synchronous motor, so the maglev system is used here to provide thrust as well as lift to the vehicle. The topology of the linear synchronous motor may be derived from that of the conventional a.c. synchronous

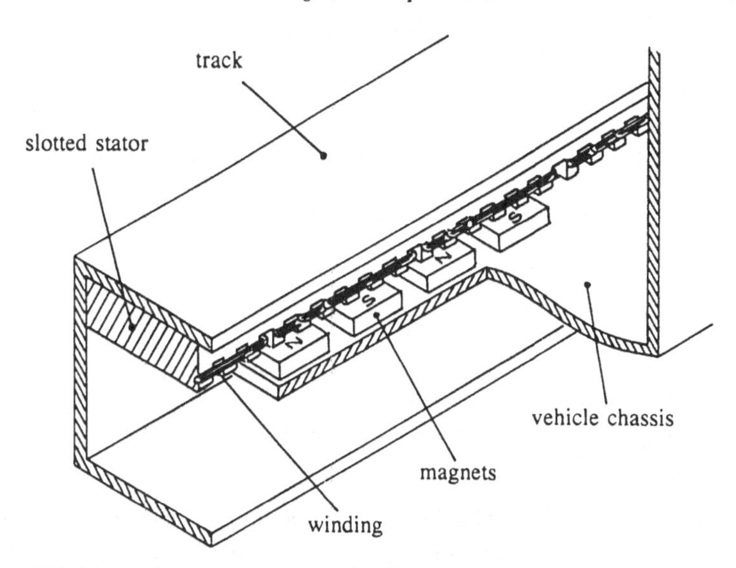

Figure 7.30. Magnetic levitation suspension for a low speed transportation system.

motor described in Section 7.4; just consider that the components in Figure 7.1(*b*) are "unrolled" and laid flat. The speed and direction of the vehicle are controlled by supplying the winding from a three-phase variable frequency inverter, and because the winding must extend along the entire track, it is usually only excited in sections as the vehicle passes.

For vehicles that run at much higher speeds, it is not possible to include wheels in the primary suspension system. In this case, the maglev components in Figure 7.30 are inverted, the track is replaced by a *conducting* material such as aluminum, and a repulsion force between the permanent magnets and induced currents in the track generates lift. While the vehicle is supported as shown in Figure 7.31, other systems are used to provide lateral stability and thrust. The lift force increases with vehicle speed, because it is the relative motion of the magnets that induces eddy currents in the aluminum, which in turn form image poles, which repel the magnets. This is no longer a magnetostatic field problem, and Equation (5.28) will not provide an applicable solution for the flux distribution, because it does not account for time-varying effects such as eddy currents. The enhanced formulation includes the track conductivity σ:

$$\nabla \times \frac{1}{\mu_0 \mu} \nabla \times \boldsymbol{A} = -\sigma \frac{\partial \boldsymbol{A}}{\partial t} + \boldsymbol{J} \tag{7.24}$$

This may still be solved by the finite element method, flux density being calculated from the potential distribution, and the forces determined.

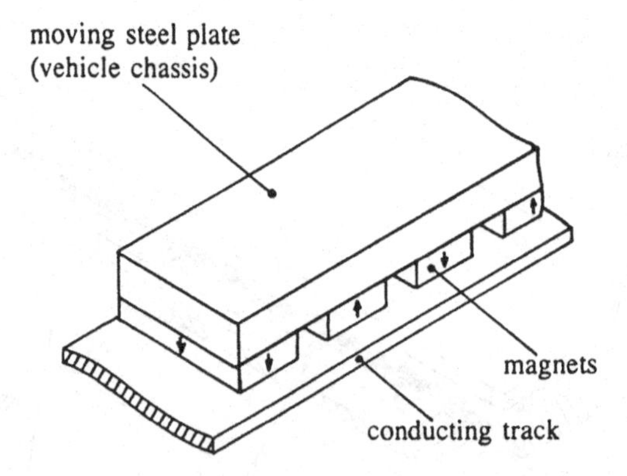

Figure 7.31. Magnetic levitation suspension for a high speed vehicle.

Alternatively, it may be solved in a simplified integral form for rectangular magnet shapes, by considering the induced eddy currents as a wake of receding image poles (Reitz, 1970). While the lift force F_L increases with velocity v as shown in Figure 7.32, the image poles also create a drag force F_D, which is directly related to F_L via

$$\frac{F_L}{F_D} = \frac{\sigma\mu\delta v}{2} \tag{7.25}$$

In this case, μ is the permeability of the track, while δ is either its thickness or its skin depth, whichever is the smaller. At high speed, the lift force approaches a constant asymptotic value, while the drag force decays as v^{-1}, continually improving the lift-to-drag ratio.

While the existence of a *drag* force is unwelcome in a linear propulsion system, it may be used to advantage in a *magnetic coupling*, where force is to be transmitted between the two components without contact. This concept is mostly used to transmit torque in rotary devices, and the two rings are usually in attraction rather than repulsion. In all the layouts considered for magnetic bearings in Figures 7.28 and 7.29, there is uniform magnetization around both of the rings, and so no torque is developed to impede the rotation. However, if the coaxial rings in Figure 7.33 are magnetized with arrays of similar numbers of *multiple* poles, then the device functions as a magnetic coupling. Two rotating fields are synchronized in the same manner as the a.c. synchronous motor described in Section 7.4, and for torque to be transmitted, there should be a phase shift δ between the fields – maximum power occurs when the torque angle $\delta = 90°$.

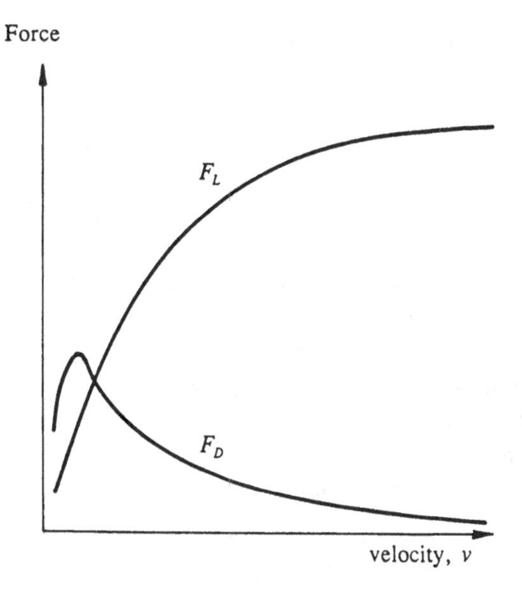

Figure 7.32. Lift and drag forces for magnets moving above a conductive track.

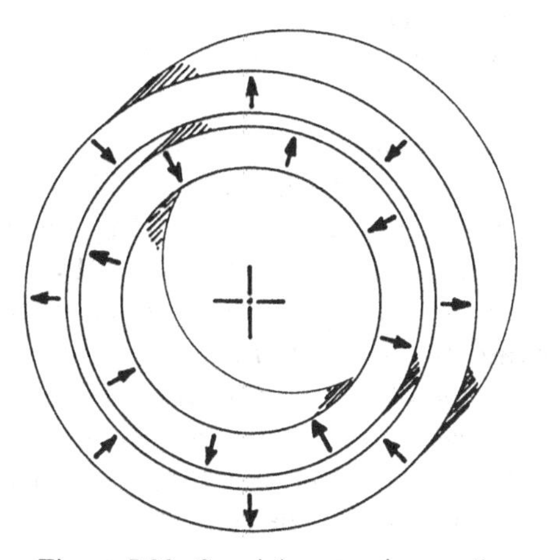

Figure 7.33. Coaxial magnetic coupling.

7.8 Sensors

The emergence of sensors as a major market for permanent magnet materials was due to broader performance requirements in the principal applications. To achieve higher efficiencies in motion control systems, many mechanical devices were replaced by electrically controlled motor drives, with examples ranging from aircraft flight controls to industrial drives to automobile anti-lock braking systems. For closed loop speed control, any opportunity for a new electric drive also presents an opportunity for a sensor, to continuously monitor angular (or linear) position or speed. With almost unlimited information processing capacity available at low cost, motion controllers became quite sophisticated, demanding better resolution and linearity of the sensors providing information to them. A variety of sensor technologies are currently employed, including silicon, optical and capacitance, in addition to magnetic. In most applications, though, it is also important for the sensor to be economical and reliable, the latter usually being synonymous with a non-contacting device to avoid such problems as contact resistance, abrasion and signal noise.

Optical sensors are the most common type found in industrial equipment, but their construction is not rugged and they function poorly in hostile environments. Because of this, magnetic sensors became widely used in automobiles, and many other applications where they would operate in dirt or fluid, at high temperatures or high vibration. Magnetic sensors are generally used to measure angular position or rotational speed, although direct linear measurement is also common. Most are pulse generators, the count being proportional to position and the frequency proportional to speed. A reed switch is a simple magnetic sensor, in which an electrical contact is formed by a moving soft iron armature, which closes under the influence of an applied field – following the operating principal of the holding force actuator. Functioning as a sensor, the change in the applied field is either implemented directly by using a multi-pole magnet or indirectly by modulating a magnet's field as with a slotted "vane". In either case, the operating frequency of the electrical contacts is limited to a low value of around 600 Hz, and the electromechanical nature of this device seldom meets today's reliability criteria.

In most magnetic sensors, the field from a permanent magnet is measured using a semiconductor element that exhibits the *Hall effect*, which was described in Section 6.4. When a current i is applied along the length of a *Hall* sensor with a characteristic constant R_H, a voltage V_H is developed across its width that is proportional to the flux density B through its thickness t:

$$V_H = \left(\frac{R_H}{t}\right) iB \tag{6.20}$$

Hall sensors can obviously be used to provide an output voltage, which is proportional to the applied field, but often their signal is processed to produce a digital output as a pulse generator – the analog signal is set to switch at predetermined "latch" and "release" levels. An important advantage of the Hall sensor is that, unlike an inductive device, its output voltage is independent of the speed of rotation, so there is no loss of information should the drive stop for any period of time.

Hall sensors are most frequently used in magnetic circuits that exhibit variable reluctance with angular position. In the very simple configuration shown in Figure 7.34, a stationary permanent magnet faces a soft iron gear wheel, whose teeth cause modulation of the magnetic circuit reluctance during rotation. The consequent variations in air gap flux are measured by a Hall sensor, mounted on the face of the magnet. Because of its simplicity, this layout achieves poor concentration of the magnet's flux into the air gap, does not optimize the change in reluctance, and is therefore limited to relatively low angular resolutions. An alternative variable reluctance device employs a slotted soft iron "vane", with either the disk or the cup topology shown in Figure 7.35. The principle of

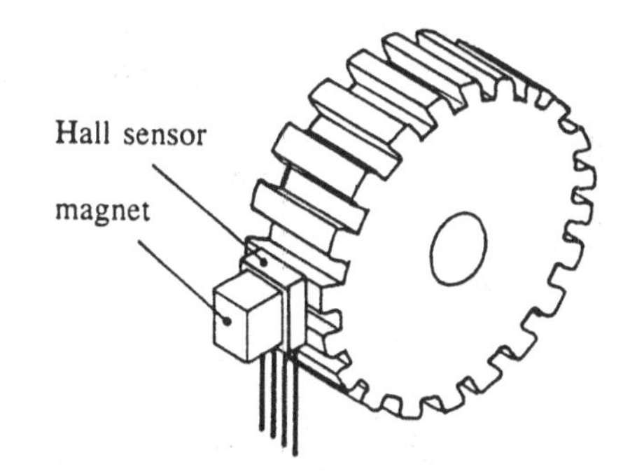

Figure 7.34. Variable reluctance magnetic sensor using a gear tooth wheel.

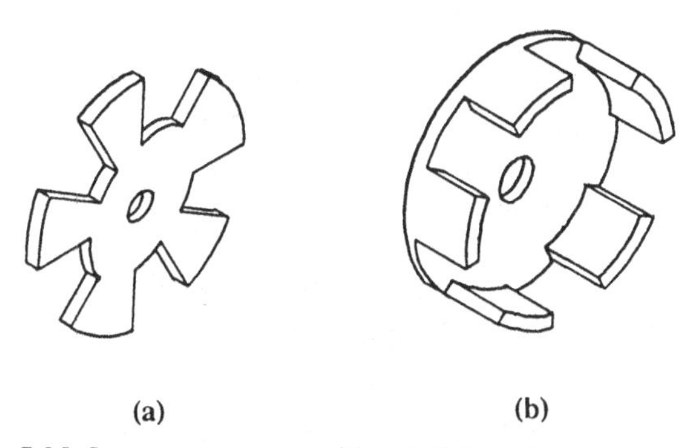

Figure 7.35. Interrupter vanes with (*a*) disk and (*b*) cup topologies.

operation using the disk configuration is illustrated in Figure 7.36, which shows how a vane is used to shunt the magnet's flux away from the Hall sensor. This layout is also restricted to relatively low angular resolutions, to achieve sufficient variation in the magnetic circuit reluctance.

Higher angular resolutions are obtained without a variable reluctance, by using a cylindrical permanent magnet rotor, made from an isotropic material with a heteropolar structure magnetized into its surface. The field distribution in such a rotor was illustrated in Figure 7.14, but for a sensor application it will more probably appear in the form of Figure 7.37. A Hall effect sensor may now be mounted directly over the circumference of a wheel, to measure the change in radial flux through its thickness as the wheel rotates. Obviously, an alternative topology can be adopted,

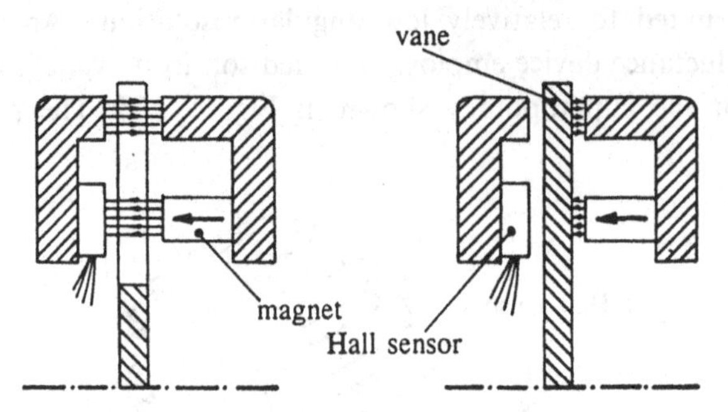

Figure 7.36. Operation of a variable reluctance magnetic sensor with a disk vane interrupter.

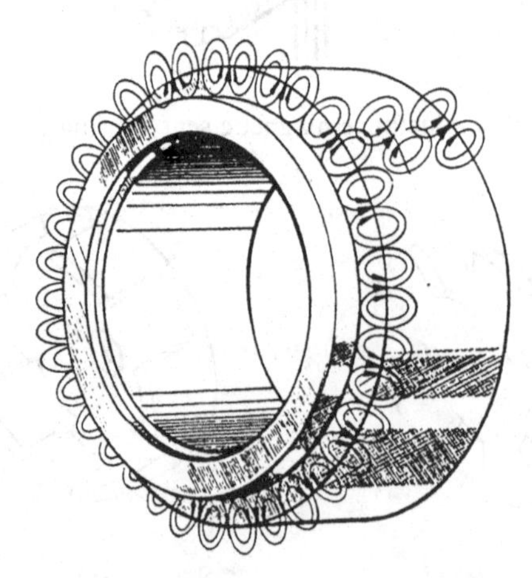

Figure 7.37. Isotropic permanent magnet wheel, with flux lines showing heteropolar structure.

with the heteropolar structure magnetized into one disk surface of the wheel, and the sensor mounted over this flat face to measure the axial flux. Because the rotor poles operate into an open magnetic circuit, the construction of this device is very simple, with only two components – a rotor magnet and a Hall sensor.

An important issue in application of magnetic sensors is the assembly tolerance between the stator and rotor components, which directly bears upon the range over which pulse generation is successfully achieved. Commercial digital Hall sensors, which incorporate signal processing circuitry to provide a pulsed output, have typical "latch" and "release" levels of around 0.03 T, the flux density that they need to experience across their thickness. Should greater sensitivity be required either to increase the range or the resolution of the unit, then the analog signal from the Hall element may be used with a custom amplifier circuit to process V_H, generated according to Equation (6.20). Alternatively, two other types of sensor may be considered, whose outputs are also independent of rotational speed but which each require only about 0.005 T for a satisfactory response. One is based upon the *Wiegand* effect in a coil, whose core is specially constructed with two dissimilar magnetic materials such that changes in the applied field cause flux jumps, and hence voltage pulses (Opie and Bossoli, 1988). The other employs a *magnetoresistive* element, whose resistance changes with magnitude and orientation of the applied field, which may be incorporated into a bridge network in an amplifier circuit to generate pulses (Campbell, 1993).

Most of the layouts described in this section are typical of magnetic sensors in that they incorporate relatively large air gaps. The greater flux densities provided by higher energy permanent magnet materials afford more reliable operation at the larger gaps associated with greater assembly tolerances. With a long air gap, the load line will place the magnet's operating point well down its demagnetization characteristic, which should preferably be substantially linear with a high coercivity. Under this condition, it was shown in Section 4.5 that the maximum air gap energy $(E_g)_{max}$ is achieved when $S=1$, the load line slope is $-\mu_0$, and the magnet's operation point is $B_m=\frac{1}{2}\mu_0 M$, $-H_m=\frac{1}{2}M$, giving

$$(E_g)_{max}=\frac{\mu_0 V_m}{2}\left(\frac{M}{2}\right)^2 \qquad (4.33)$$

This expression serves to show that, by increasing the volume or energy density of the magnet material, the corresponding rise in air gap energy can be utilized to provide a greater range for the sensing element. When a magnetic sensor employs a single permanent magnet in a simple variable reluctance configuration such as those shown in Figures 7.34 and 7.36, it is common to find a small magnet of the highest energy density available, such as fully dense $Nd_2Fe_{14}B$. However, the more complex heteropolar structure of Figure 7.37 provides very poor utilization of a large magnet volume, at best only one pair of adjacent poles interacting with the sensor element at any time; sintered or bonded ceramic ferrite magnets are most common in this situation, and usually an isotropic grade to facilitate magnetization of the pole array.

7.9 Steady fields

There are several applications for permanent magnets that demand a very steady magnetic field in a cavity. One that also requires a very uniform field is *magnetic resonance imaging* (MRI). MRI is a technique for medical diagnosis of the human body, and also for animal inspection in the food industry. To create an image, an MRI scanner generates a strong magnetic field through the body region. This field attracts hydrogen protons, aligning them parallel to the field. Radio waves are used to scatter the protons, and when these waves are switched off, realignment of the protons with the field generates their own radio waves, which are detected by sensors. The data are used to construct three-dimensional images of organs and tissue by assembling parallel scans, which show "slices" of the body region. Because protons are plentiful in fat and water but absent in bones, MRI scanners can see through bones to provide unobstructed views of internal organs, such as the brain.

In early MRI scanners, good image quality was only achieved with very high fields up to 1.5 T in the cavity, which required the use of superconducting coils. Later advances in scanner technology have allowed satisfactory images to be obtained at much lower fields in the range 0.1–0.5 T, depending on the size of the unit. Furthermore, not all scanners have large cavities to accommodate a whole human body – some are designed for specialized studies, such as head, arm and leg imaging. These developments have made it possible to use permanent magnets as the field source, thus eliminating the need for cryogenic hardware and a power supply. Because of the large cavity, only high coercivity magnets are employed, though units have been built with ceramic ferrites as well as sintered Sm_2Co_{17} and $Nd_2Fe_{14}B$. Many novel magnet configurations have been investigated for their ability to produce a steady, uniform field in the cavity, some of which require very difficult assembly of many large permanent magnet blocks, each pre-magnetized with different orientations. One practical layout shown in Figure 7.38 employs mechanical adjustment of spacing to modify the field in the cavity (Miyamoto *et al.*, 1989). Working in conjunction with the permanent magnets, the pole pieces in this unit are designed to tailor the flux distribution in the cavity. The only way to calculate the flux with sufficient accuracy is to employ magnetic potential again in the form of Equation (5.28), as

$$\nabla \times \frac{1}{\mu_0 \mu} \nabla \times \boldsymbol{A} = \nabla \times \frac{\boldsymbol{M}_r}{\mu} \tag{7.26}$$

This may be solved by the finite difference or finite element methods, the flux density then being calculated from the potential distribution. Because the air gap is so large, this must be treated as a three-dimensional problem, but to avoid the great complexity of this procedure, axi-symmetry can frequently be used to reduce it to two dimensions.

The steady fields of permanent magnets are commonly used to focus electron beams in *traveling wave tubes* (TWTs), such as microwave or millimeter-wave generators and amplifiers. An axial field in the cylindrical cavity is needed to constrain electrons to travel in a narrow beam along the length of the tube. Early TWTs used a uniform axial field along the entire cavity, as provided by a long solenoid electromagnet. The power supply is eliminated by replacing the solenoid with a large cylindrical alnico magnet, which also has a uniform axial field in its bore. Both these

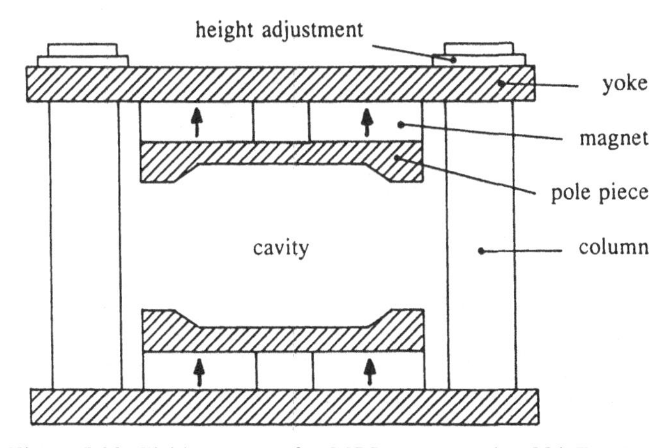

Figure 7.38. Field structure for MRI scanner using $Nd_2Fe_{14}B$.

units are bulky and heavy. It was later found that the field need not be uniform to focus the electron beam, but should only be oriented along the axis of the tube. This requirement is satisfied with a steady, periodic field using a linear magnet array, which is shown in cross-section in Figure 7.39. For comparable beam focussing performance, dividing the magnet into an array provides a significant reduction in size and weight, even with alnico magnets, but especially so with high energy rare earth materials. In this application, it is important that the field be very steady, so the permanent magnets must have good temperature stability. As has been discussed in Section 3.4, the reversible temperature coefficient is significantly reduced by adjusting the compositions of $SmCo_5$ and Sm_2Co_{17}, though the improvement is accompanied by a decline in the magnet's energy product. For instance, Sm_2Co_{17} with $(BH)_{max}$ of 240 kJ/m^3 has a coefficient of $-0.030\%/°C$, but when the composition is varied to improve this to $-0.010\%/°C$, there is a reduction in $(BH)_{max}$ to 140 kJ/m^3 (Table 3.1).

If the periodic field is not parallel to the bore's axis, as with the magnet structure of Figure 7.40, then an electron or ion beam can be forced into a sinusoidal path, which causes it to emit a very intense, focussed beam of gamma radiation. Such "*wigglers*" are the most powerful generators of X-rays and ultraviolet radiation. Figure 7.40 shows just one possible construction for a wiggler, there being several other ways to assemble an array of permanent magnets to produce a spatially periodic field.

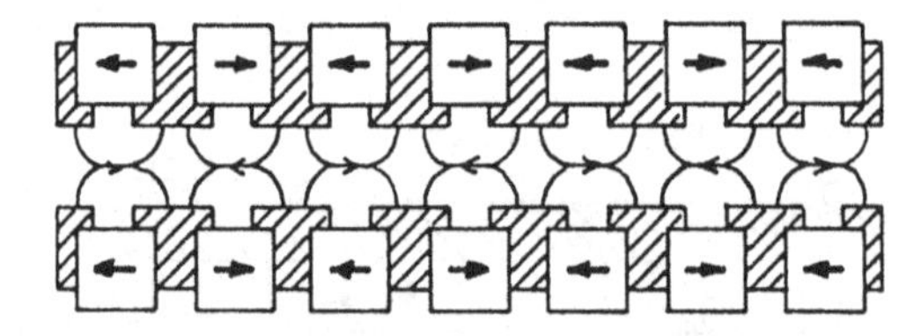

Figure 7.39. Periodic magnet structure for beam focussing in a traveling wave tube.

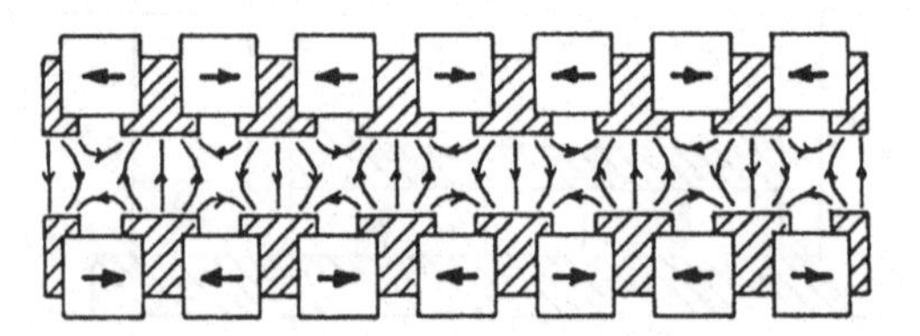

Figure 7.40. Periodic magnet structure for generating gamma radiation.

References

Campbell, P. (1974). Principles of a permanent magnet axial field d.c. machine. *IEE Proceedings*, **121**, 1489–94.

Campbell, P. (1990). A new Nd–Fe–B brushless d.c. motor for a digital data storage spindle drive. *11th International Workshop on Rare-Earth Magnets and their Applications*, pp. 221–30. Pittsburgh: Carnegie Mellon University.

Campbell, P. (1993). Magnetoresistive sensors for high resolution position encoding. *Sensors and Actuators 1993* (SP-948), pp. 25–33. Warrendale: Society of Automotive Engineers, Inc.

Campbell, P., Tsals, I. and Matsuura, D. (1986). Apparatus and method for controlling the parenteral administration of fluids. United States Patent 4,626,241.

Earnshaw, S. (1839). On the nature of the molecular forces which regulate the constitution of the luminiferous ether. *Transactions of the Cambridge Philosophical Society*, **7**, 97–112.

Howe, D., Birch, T. S. and Gray, P. (1987). The potential for Nd–Fe–B in electrical machines. *9th International Workshop on Rare-Earth Magnets and their Applications*, pp. 65–83. Bad Hoffen: Deutsche Physikalische Gesellschaft.

Kenjo, T. and Nagamori, S. (1985). *Permanent Magnet and Brushless DC Motors*, Oxford: Clarendon Press.

Leenhouts, A. C. (1987). *The Art and Practice of Step Motor Control*, Ventura: Intertec Communications, Inc.

Miller, T. J. E. (1989). *Brushless Permanent-Magnet and Reluctance Motor Drives*, Oxford: Clarendon Press.

Miyamoto, T., Sakurai, H., Takabayashi, H. and Aoki, M. (1989). A development of a permanent magnet assembly for MRI devices using Nd–Fe–B material. *IEEE Transactions on Magnetics*, **25**, 3907–9.

Opie, J. E. and Bossoli, J. W. (1988). A new era of application for the Wiegand effect. *Sensors and Actuators 1988* (SP-737), pp. 1–6. Warrendale: Society of Automotive Engineers, Inc.

Reitz, J. R. (1970). Forces on moving magnets due to eddy currents. *Journal of Applied Physics*, **41**, 2067–71.

Say, M. G. and Taylor, E. O. (1980). *Direct Current Machines*, Bath: The Pitman Press.

Wagner, J. A. (1982). The shorted turn in the linear actuator of a high performance disk drive. *IEEE Transactions on Magnetics*, **18**, 1770–2.

Yonnet, J.-P. (1981). Permanent magnet bearings and couplings. *IEEE Transactions on Magnetics*, **17**, 1169–73.

Index